Dimensions Math
Workbook 4B

Authors and Reviewers
Jenny Kempe
Bill Jackson
Tricia Salerno
Allison Coates
Cassandra Turner

Singapore Math Inc.

Published by Singapore Math Inc.

19535 SW 129th Avenue
Tualatin, OR 97062
www.singaporemath.com

Dimensions Math® Workbook 4B
ISBN 978-1-947226-25-8

First published 2019
Reprinted 2020 (twice), 2021 (three times), 2023 (twice)

Copyright © 2019 by Singapore Math Inc.
All rights reserved. This book or any portion thereof may not be reproduced or used in any manner whatsoever without the express written permission of the publisher.

Printed in China

Acknowledgments

Design and illustration by Cameron Wray with Carli Bartlett.

Contents

Chapter	Exercise	Page
Chapter 10 **Measurement**	Exercise 1	1
	Exercise 2	4
	Exercise 3	8
	Exercise 4	11
	Exercise 5	14
	Exercise 6	17
	Exercise 7	20
	Exercise 8	23
	Exercise 9	26
Chapter 11 **Area and Perimeter**	Exercise 1	29
	Exercise 2	32
	Exercise 3	35
	Exercise 4	39
	Exercise 5	42
	Exercise 6	45

Chapter	Exercise	Page
Chapter 12 **Decimals**	Exercise 1	49
	Exercise 2	52
	Exercise 3	56
	Exercise 4	59
	Exercise 5	63
	Exercise 6	66
	Exercise 7	68
	Exercise 8	70
	Exercise 9	73
	Exercise 10	76
Chapter 13 **Addition and Subtraction of Decimals**	Exercise 1	79
	Exercise 2	81
	Exercise 3	83
	Exercise 4	86
	Exercise 5	88
	Exercise 6	90
	Exercise 7	92
	Exercise 8	95
	Exercise 9	98
	Exercise 10	101

Chapter	Exercise	Page
Chapter 14 **Multiplication and Division of Decimals**	Exercise 1	107
	Exercise 2	109
	Exercise 3	112
	Exercise 4	115
	Exercise 5	117
	Exercise 6	119
	Exercise 7	122
	Exercise 8	125
	Exercise 9	127
Chapter 15 **Angles**	Exercise 1	131
	Exercise 2	135
	Exercise 3	139
	Exercise 4	144
	Exercise 5	147
	Exercise 6	151
Chapter 16 **Lines and Shapes**	Exercise 1	157
	Exercise 2	160
	Exercise 3	162
	Exercise 4	166
	Exercise 5	170
	Exercise 6	174
	Exercise 7	177

Chapter	Exercise	Page
Chapter 17 **Properties of Cuboids**	Exercise 1	183
	Exercise 2	186
	Exercise 3	190
	Exercise 4	194
	Exercise 5	198
	Exercise 6	204

This workbook includes **Basics**, **Practice**, **Challenge**, and **Check** sections to review and deepen math skills.

Chapter 10 Measurement

Exercise 1

Basics

1 (a) 6 m = 6 × 100 cm = [ ] cm

(b) 6 m 52 cm = [ ] cm + 52 cm

= [ ] cm

| 1 cm = 10 mm |
| 1 m = 100 cm |
| 1 km = 1,000 m |
| 1 kg = 1,000 g |
| 1 L = 1,000 mL |

2 (a) 5,000 mL = [ ] L

(b) 5,450 mL = [ ] L [ ] mL

3 5,450 cm = [ ] m [ ] cm

4 4 L 450 mL $\xrightarrow{+\ 3\ L}$ [ ] L 450 mL $\xrightarrow{+\ 800\ mL}$ [ ] L [ ] mL

4 L 450 mL + 3 L 800 mL = [ ] L [ ] mL

5 9 kg 600 g $\xrightarrow{-\ 5\ kg}$ [ ] kg 600 g $\xrightarrow{-\ 450\ g}$ [ ] kg [ ] g

9 kg 600 g − 5 kg 450 g = [ ] kg [ ] g

10-1 Metric Units of Measurement

1

6 (a) 3 × 8 m = ☐ m

(b) 3 × 42 cm = ☐ cm = ☐ m ☐ cm

(c) 3 × 8 m 42 cm = ☐ m ☐ cm

Practice

7 (a) 1 km − 300 m = ☐ m   (b) 1 km − 30 m = ☐ m

(c) 1 m − 3 cm = ☐ cm   (d) 1 cm − 3 mm = ☐ mm

8 A bar of soap weighs 235 g. How much do 8 bars of soap weigh? Express the answer in compound units.

9 2 blocks of concrete weigh 45 kg altogether. One of the blocks weighs 35 kg 230 g. How much does the other block weigh? Express the answer in compound units.

10 A rope is 12 m 40 cm long. 3 pieces that are each 2 m 85 cm long are cut from it. How long is the remaining piece of rope? Express the answer in compound units.

11 Amy bought 5 bottles of juice. Each bottle contains 650 mL of juice. How much juice does she have in all? Express the answer in compound units.

Exercise 2

Basics

1 The diagram below shows the relative lengths of some customary units of length. Use the diagram to find the missing numbers.

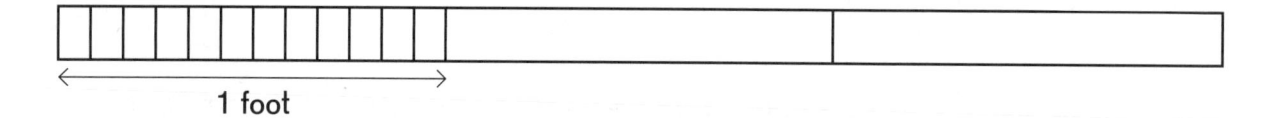

1 inch

1 foot

1 yard

(a) 1 ft = [ ] in

(b) 1 yd = [ ] ft

(c) 1 yd = [ ] in

2 (a) 4 ft = 4 × 12 in = [ ] in

(b) 4 ft 8 in = [ ] in + 8 in = [ ] in

3 (a) 5 ft = [ ] in

(b) 64 in = [ ] in + 4 in = [ ] ft [ ] in

4   10-2 Customary Units of Length

4. (a) 47 ÷ 3 is [ ] with a remainder of [ ].

   (b) 47 ft = [ ] yd [ ] ft

5. (a) 7 in + [ ] in = 1 ft

   (b) 7 in + 8 in = [ ] ft [ ] in

   (c) 1 ft − 2 in = [ ] in

   (d) 2 ft − 2 in = [ ] ft [ ] in

   (e) 2 ft 3 in − 5 in = [ ] ft [ ] in

6. 4 ft 7 in —+ 6 ft→ [ ] ft 7 in —+ 8 in→ [ ] ft [ ] in

   4 ft 7 in + 6 ft 8 in = [ ] ft [ ] in

7. 9 ft 3 in —− 7 ft→ [ ] ft 3 in —− 5 in→ [ ] ft [ ] in

   9 ft 3 in − 7 ft 5 in = [ ] ft [ ] in

Practice

8 (a) Fill in the missing numbers.

Object	Measurement	
length of a couch	6 ft 2 in	☐ in
height of a couch	☐ ft ☐ in	40 in
length of a bookshelf	2 yd 5 in	☐ in
height of a bookshelf	☐ ft ☐ in	27 in

(b) Find the sum of the lengths of the bookshelf and the couch. Express the answer in feet and inches.

(c) Find the difference in height of the bookshelf and the couch. Express the answer in feet and inches.

9 A bolt (roll) of fabric has 40 yards of fabric. 2 yd 2 ft were cut from it to make a dress and 1 yd 2 ft to make a skirt.

(a) What is the length of fabric that was used? Express the answer in yards and feet.

(b) What is the length of fabric that was left on the bolt? Express the answer in yards and feet.

Challenge

10 1 mile is equal to 5,280 feet. 4 laps around a track is 1 mile.

(a) What is the length in feet of one lap around this track?

(b) What is the length in yards of one lap around this track?

Exercise 3

Basics

1 The picture shows a scale marked in pounds. The interval between each tick mark is 1 ounce. You can use it to answer the following questions.

(a) 1 lb = ☐ oz

(b) 3 lb = 3 × ☐ oz = ☐ oz

3 lb 7 oz = ☐ oz + 7 oz = ☐ oz

The block weighs ☐ oz.

(c) 2 lb = ☐ oz

36 oz = ☐ oz + 4 oz = ☐ lb ☐ oz

(d) 1 lb − 9 oz = ☐ oz

4 lb − 9 oz = ☐ lb ☐ oz

2 3 lb 7 oz $\xrightarrow{+\,2\text{ lb}}$ ☐ lb 7 oz $\xrightarrow{+\,9\text{ oz}}$ ☐ lb ☐ oz

3 lb 7 oz + 2 lb 9 oz = ☐ lb ☐ oz

3 6 lb 2 oz $\xrightarrow{-\,4\text{ lb}}$ ☐ lb 2 oz $\xrightarrow{-\,9\text{ oz}}$ ☐ lb ☐ oz

6 lb 2 oz − 4 lb 9 oz = ☐ lb ☐ oz

10-3 Customary Units of Weight

Practice

4 (a) Fill in the missing numbers.

Object	Weight	
a boot	☐ lb ☐ oz	23 oz
a bottle of ketchup	3 lb 14 oz	☐ oz
a pumpkin	4 lb 2 oz	☐ oz
a book	☐ lb ☐ oz	33 oz

(b) Find the sum of the weights of the lightest and heaviest objects. Express the answer in compound units.

(c) Find the difference in the weights of the lightest and heaviest objects. Express the answer in compound units.

5. Alex weighed 6 lb 4 oz at birth. Emma weighed 120 oz. Who weighed more and by how much? Express the answer in compound units.

6. A brick weighs 5 lb 10 oz. How much do 3 bricks weigh? Express the answer in compound units.

Challenge

7. Dion found that 100 pennies weigh 9 ounces and 100 dimes weigh 8 ounces. How much do 500 pennies and 500 dimes weigh altogether? Express the answer in compound units.

Exercise 4

Basics

1 The diagram below shows the relative sizes of some customary units of capacity. Use the diagram to find the missing numbers.

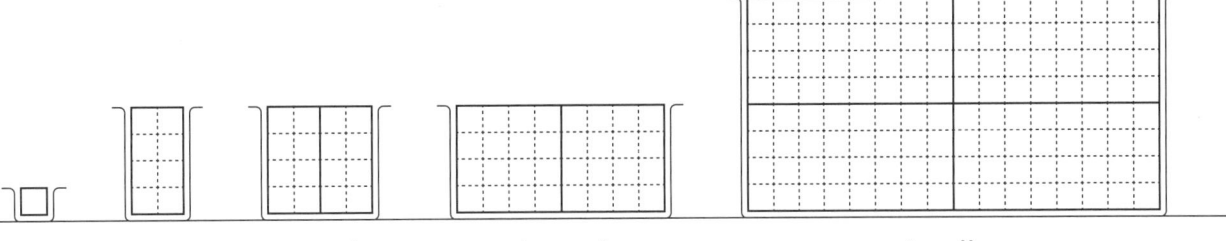

1 fluid ounce   1 cup   1 pint   1 quart   1 gallon

(a) 1 c = [ ] fl oz

(b) 1 pt = [ ] c   1 pt = [ ] fl oz

(c) 1 qt = [ ] pt   1 qt = [ ] c

(d) 1 gal = [ ] qt   1 gal = [ ] pt

(e) 1 gal = [ ] c   1 gal = [ ] fl oz

2 (a) 3 gal = 3 × [ ] qt = [ ] qt

(b) 3 gal 3 qt = [ ] qt + 3 qt = [ ] qt

3 (a) 62 ÷ 4 is [ ] with a remainder of [ ].

(b) 62 qt = [ ] gal [ ] qt

Practice

④ (a) 5 gal = ☐ qt   (b) 5 qt = ☐ c

(c) 5 pt = ☐ c   (d) 5 c = ☐ fl oz

⑤ (a) 32 fl oz = ☐ c   (b) 32 c = ☐ pt

(c) 32 pt = ☐ qt   (d) 32 qt = ☐ gal

⑥ (a) 7 c = ☐ pt ☐ c

(b) 7 c = ☐ qt ☐ c

(c) 15 qt = ☐ gal ☐ qt

(d) 30 fl oz = ☐ c ☐ fl oz

⑦ A recipe calls for 3 tablespoons of soy sauce. Yara only has a teaspoon. She knows there are 3 teaspoons in a tablespoon. She is doubling the recipe. How many teaspoons of soy sauce does she need?

8 A coffee shop sells coffee in 3 sizes: redwood, cedar, and fir. A redwood is 16 fl oz. A cedar is 14 fl oz. A fir is 10 fl oz.

(a) Which sizes are more than 1 cup?

(b) What is the total capacity of 2 cedars? Express the answer in cups and fluid ounces.

(c) A family orders 2 redwoods, 1 cedar, and 3 firs. What is the total amount of coffee they got? Express the answer in cups and fluid ounces.

Exercise 5

Basics

1 (a) 4 min = 4 × ☐ s = ☐ s

(b) 4 min 18 s = ☐ s + 18 s = ☐ s

> 1 min = 60 s
> 1 h = 60 min
> 1 day = 24 h
> 1 week = 7 days

2 (a) 3 h = ☐ min

(b) 190 min = ☐ min + 10 min = ☐ h ☐ min

3 (a) 5 × 3 h = ☐ h

(b) 5 × 35 min = ☐ h ☐ min

(c) 5 × 3 h 35 min = ☐ h ☐ min

4 (a) 1 h − 34 min = ☐ min

(b) 5 h − 34 min = ☐ h ☐ min

5 3 min 35 s —+ 8 min→ ☐ min 35 s —+ 48 s→ ☐ min ☐ s

3 min 35 s + 8 min 48 s = ☐ min ☐ s

6 7 h 20 min —− 2 h→ ☐ h 20 min —− 34 min→ ☐ h ☐ min

7 h 20 min − 2 h 34 min = ☐ h ☐ min

Practice

7 (a) 95 min = ☐ h ☐ min

(b) What time is it 95 min after 11:35 a.m.?

(c) What time is it 95 min before 11:35 a.m.?

8 How many hours is it from Monday at 9:00 a.m. to Friday at 5:00 p.m. of the same week?

9 Jacob used a stopwatch to time two races. After the first race, the stopwatch said 00:03:42. After the second race, it said 00:07:27. Jacob realized that he forgot to reset the stopwatch back to 0 after the first race. How long was the second race in minutes and seconds?

10 Laila went on a bus ride that lasted 3 h 14 min. Gavin went on a train ride that lasted exactly half as long as Laila's bus ride. How long was Gavin's train ride in hours and minutes?

Challenge

11 There are 365 days in a year, and 366 days in a leap year. A leap year occurs every 4 years. There are 10 years in a decade. What is the least possible number of days in a decade?

Exercise 6

Check

1 (a) 5 cm 2 mm = ☐ mm

(b) 5 min 5 s = ☐ s

(c) 5 lb 2 oz = ☐ oz

(d) 5 kg 2 g = ☐ g

(e) 26 in = ☐ ft ☐ in

(f) 35 qt = ☐ gal ☐ qt

(g) 6 km 735 m + 2 km 825 m = ☐ km ☐ m

(h) 6 lb 4 oz − 2 lb 9 oz = ☐ lb ☐ oz

(i) 10 ft 7 in − 4 ft 9 in = ☐ ft ☐ in

2 A soaker hose is 200 ft long. Brandon needs it to reach a distance of 58 yd. How many extra yards and feet of hose will he have?

3 Brandon has an automatic timer for his soaker hose. He wants to run the water for 200 minutes and have it shut off at 7:30 a.m. What time should he set the timer to start?

4 2 L 658 mL of Solution A and 1 L 162 mL of Solution B are mixed together, and then 500 mL is poured out. How much of the mixture is left? Express the answer in liters and milliliters.

Challenge

5 3 c of orange juice, 1 pt of peach juice, and 1 c 6 fl oz of mango juice are mixed together and poured equally into 6 cups. How many fluid ounces of juice are in each cup?

6 A slime mold moves 120 mm every minute.

(a) How far will it move in 15 seconds?

(b) How many centimeters will it move in 5 minutes?

7 A "hand" is used to measure the height of horses. 1 hand equals 4 inches. If a horse is 14 hands tall, how tall is it in feet and inches?

Exercise 7

Basics

1 (a) 1 ft = ☐ in

(b) $\frac{1}{4}$ ft = $\frac{1}{4}$ × ☐ in = ☐ in

(c) $\frac{3}{4}$ ft = $\frac{3}{4}$ × ☐ in = ☐ in

2 (a) 1 min = ☐ s

(b) $\frac{1}{3}$ min = $\frac{1}{3}$ × ☐ s = ☐ s

(c) $\frac{1}{10}$ min = $\frac{1}{10}$ × ☐ s = ☐ s

(d) $\frac{5}{6}$ min = $\frac{5}{6}$ × ☐ s = ☐ s

3 Express 8 ounces as a fraction of a pound.

☐ oz = 1 lb

8 oz = $\frac{8}{\square}$ lb = $\frac{1}{\square}$ lb

4 Express 40 cm as a fraction of 1 m.

☐ cm = 1 m

40 cm = $\frac{40}{\square}$ m = $\frac{2}{\square}$ m

10-7 Fractions and Measurement — Part 1

Practice

5 (a) $\frac{3}{4}$ lb = ☐ oz

(b) $\frac{3}{4}$ qt = ☐ c

(c) $\frac{5}{12}$ h = ☐ min

(d) $\frac{5}{12}$ day = ☐ h

(e) $\frac{3}{5}$ m = ☐ cm

(f) $\frac{3}{5}$ L = ☐ mL

6 Give the answers in simplest form.

(a) 40 min = ☐/☐ h

(b) 8 in = ☐/☐ ft

(c) 450 m = ☐/☐ km

(d) 16 cm = ☐/☐ m

7 Two identical cans weigh $\frac{7}{8}$ lb together.

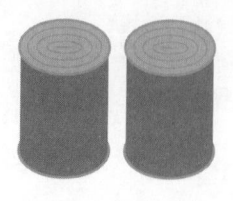

(a) How many ounces do both cans weigh together?

(b) How many ounces does one can weigh?

8 John took $\frac{7}{12}$ of an hour to wash his car. He spent $\frac{2}{5}$ of the time it took him to wash the car cleaning the interior.

(a) How many minutes did it take him to wash the car?

(b) How many minutes did it take him to clean the interior?

9 A block of cheese weighed 4 lb. $\frac{5}{8}$ of it was used to make macaroni and cheese. How much does the block of cheese weigh now? Express the answer in compound units.

Exercise 8

Basics

1 (a) $\frac{3}{4}$ ft = $\frac{3}{4}$ × 12 in = ☐ in

(b) $1\frac{3}{4}$ ft = 1 ft ☐ in

(c) $1\frac{3}{4}$ ft = 12 in + ☐ in = ☐ in

2 (a) 3 h = ☐ min

(b) $\frac{2}{3}$ h = ☐ min

(c) $3\frac{2}{3}$ h = ☐ min

3 Express 25 cm as a fraction of $1\frac{1}{2}$ m.

$1\frac{1}{2}$ m = ☐ cm

$\frac{25}{\Box} = \frac{1}{\Box}$

25 cm is $\frac{\Box}{\Box}$ of $1\frac{1}{2}$ m.

4 Express 24 h as a fraction of $1\frac{1}{2}$ days.

$1\frac{1}{2}$ days = ☐ h

$\frac{24}{\Box} = \frac{2}{\Box}$

24 h is $\frac{\Box}{\Box}$ of $1\frac{1}{2}$ days.

Practice

5 (a) $2\frac{7}{10}$ km = ☐ km ☐ m

(b) $3\frac{3}{4}$ lb = ☐ lb ☐ oz

6 (a) $4\frac{3}{4}$ ft = ☐ in

(b) $3\frac{3}{5}$ min = ☐ s

(c) $6\frac{2}{5}$ cm = ☐ mm

(d) $5\frac{5}{20}$ m = ☐ cm

(e) $2\frac{1}{2}$ c = ☐ fl oz

(f) $3\frac{2}{3}$ yd = ☐ ft

7 Carlos ran $1\frac{3}{5}$ km each day three days in a row. How far did he run altogether? Express the answer in kilometers and meters.

8 Two boards laid end-to-end measure $2\frac{1}{3}$ ft. How long are 5 such boards laid end-to-end? Express the answer in feet and inches.

9 A bag of beans weighed $2\frac{1}{2}$ lb. 12 ounces of beans were used for soup.

(a) What fraction of the bag of beans was used?

(b) Another $1\frac{1}{8}$ lb of beans was used to make chili. How many ounces does the bag of beans now weigh?

Exercise 9

Check

1 1 inch is about $2\frac{1}{2}$ cm. Using this approximation, about how many centimeters is 1 foot?

2 Laura swam a mile in $20\frac{2}{3}$ min. Express this time in minutes and seconds.

3 Dennis is 75 inches tall. Express his height in feet using a mixed number in simplest form.

4 Aliyah has $2. She spent 80¢ on a snack. What fraction of her money did she spend?

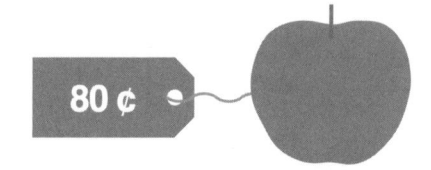

5 6 hours is what fraction of $2\frac{1}{2}$ days? Give the answer in simplest form.

6 Irene had 5 lb of flour. She used $1\frac{1}{2}$ lb of flour on Monday, $\frac{3}{4}$ lb of flour on Tuesday, and $2\frac{1}{8}$ lb of flour on Wednesday.

(a) How much flour did she use all three days? Give the answer in pounds and ounces.

(b) How many ounces of flour does she have left?

7 Dexter had two pieces of fabric. One was $2\frac{2}{3}$ yd long and the other was $3\frac{2}{3}$ yd long. He used $5\frac{1}{3}$ yd of fabric for a project. How many inches of fabric does he have left over?

8 A stack of 200 quarters is 35 cm tall. How many millimeters thick is one quarter? Give the answer as a mixed number in simplest form.

Challenge

9 In the U.S., temperature is measured in degrees Fahrenheit (°F). In most other countries, temperature is measured in degrees Celsius (°C). To convert °C to °F, multiply the numbers of degrees Celsius by $\frac{9}{5}$ and then add 32.

(a) Water freezes at 0°C. What temperature is that in °F?

(b) Water boils at 100°C. What temperature is that in °F?

(c) What is 20°C in °F?

(d) Normal body temperature is 37°C. What is this in °F? Express the answer as a mixed number in simplest form.

(e) On Thursday, a weather station said it was 25°C in Vancouver, Canada. The next day, a U.S. weather station said it was 62°F there. On which day was it cooler and by how many degrees Fahrenheit?

Chapter 11 Area and Perimeter

Exercise 1

Basics

1 Express the area of the rectangle in square inches.

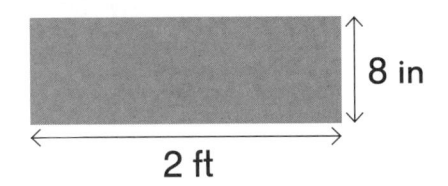

2 × 12 in = 24 in

24 × 8 = ☐

Area: ☐ in²

2 Express the area of the rectangle in square centimeters.

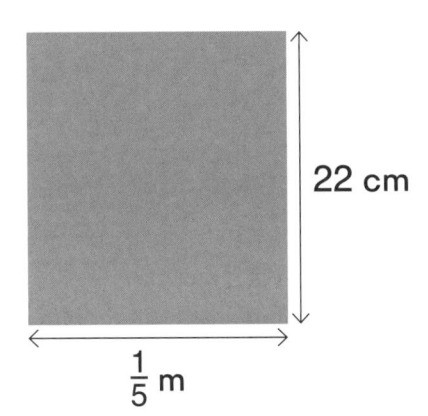

$\frac{1}{5}$ × 100 cm = ☐ cm

☐ × 22 = ☐

Area: ☐ cm²

3 Express the area of the square in square feet.

1 yd 1 ft = 4 ft

4 × 4 = ☐

Area: ☐ ft²

11-1 Area of Rectangles — Part 1

Practice

4 Express the area of each rectangle in square centimeters.

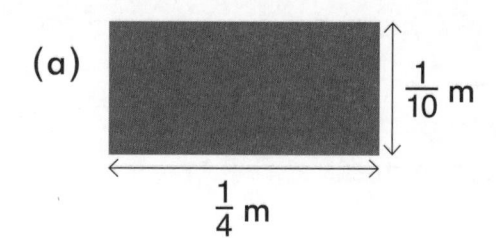

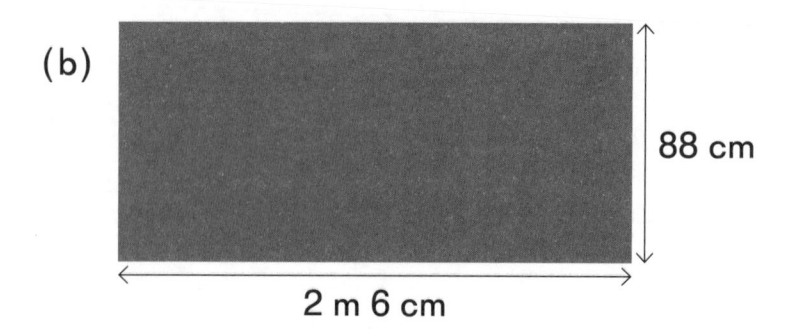

(a) $\frac{1}{10}$ m, $\frac{1}{4}$ m

(b) 88 cm, 2 m 6 cm

5 A rectangular piece of cardboard is $1\frac{1}{3}$ ft long. It is half as wide as it is long.

(a) Express the length in inches.

(b) Express the area in square inches.

6 A room is 3 yards long and $2\frac{2}{3}$ yd wide. How many 1-ft square tiles are needed to cover the floor?

Challenge

7 A rectangular piece of cardboard is 3 ft 11 in long and 2 ft 4 in wide. What is the maximum number of 5 in by 8 in rectangles that can be cut from it?

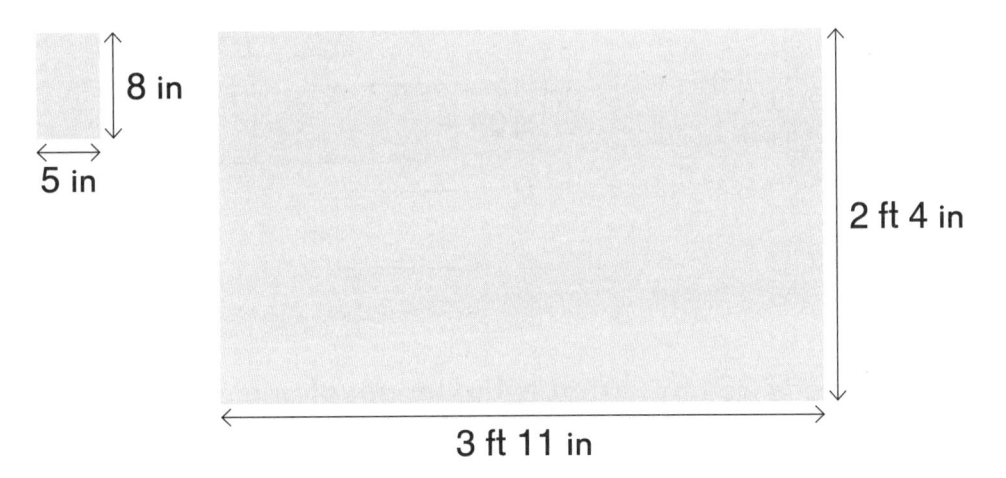

Exercise 2

Basics

1 The area of a rectangle is 126 cm². One side is 9 cm long. How long is the other side?

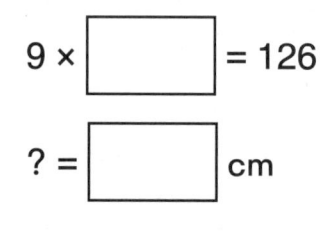

126 cm²   9 cm
?

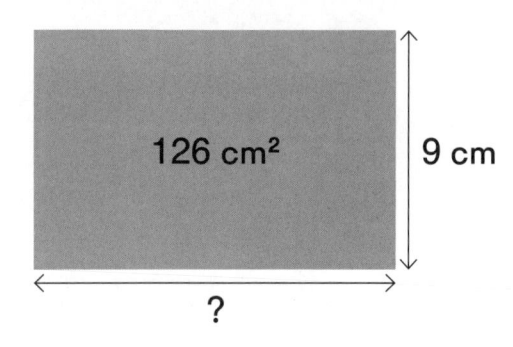

9 × ☐ = 126

? = ☐ cm

2 The area of a rectangle is 104 in². One side is $\frac{2}{3}$ ft long. How long is the other side in inches?

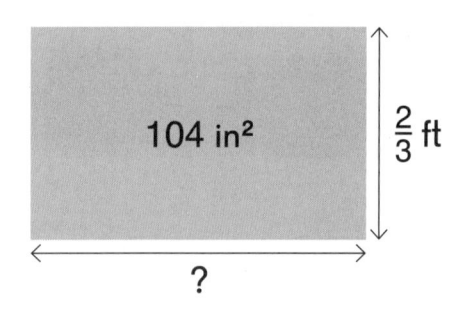

104 in² $\frac{2}{3}$ ft
?

$\frac{2}{3}$ × 12 in = ☐ in

104 ÷ ☐ = ☐

? = ☐ in

3 The area of a square is 25 cm². What is the length of one side?

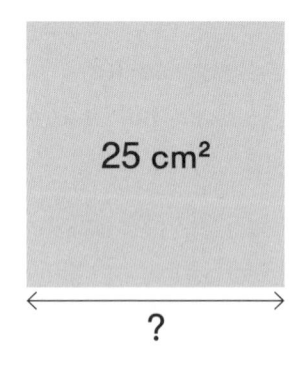

25 cm²
?

? × ? = 25

? = ☐ cm

32   11-2 Area of Rectangles — Part 2

Practice

4 The area of a rectangle is 112 m². One side is 8 m long. How long is the other side?

5 Both rectangles below have the same area. What is the unknown length? Express the answer in inches.

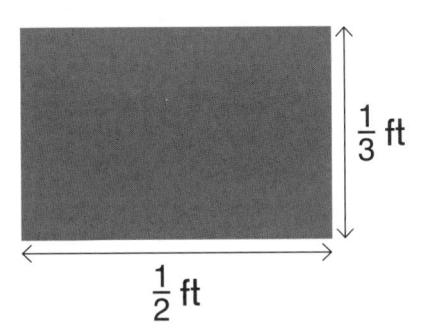

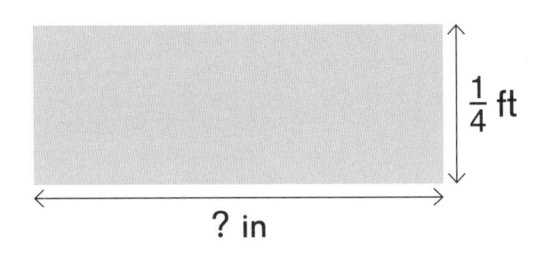

$\frac{1}{2}$ ft, $\frac{1}{3}$ ft

? in, $\frac{1}{4}$ ft

6 It costs $8 per square foot to carpet a room. The total cost to carpet the room is $336. If one side of the room is 6 ft long, what is the length of the other side of the room?

Challenge

7 This figure is made up of 8 squares. The total area of this figure is 288 cm². What is the length of the side of one square?

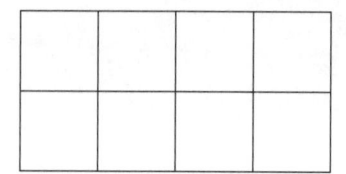

8 Both rectangles below have the same area. What is the unknown length? Express the answer in centimeters and millimeters.

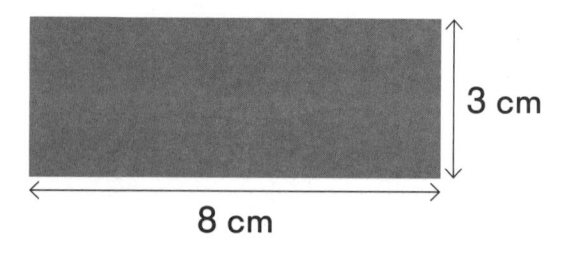

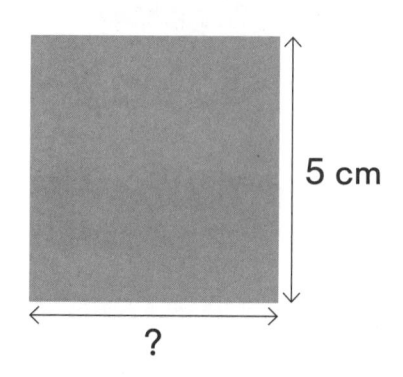

3 cm
8 cm

5 cm
?

Exercise 3

Basics

1 There are 3 identical pieces of paper. A square corner is cut from the first piece of paper.

(a) Complete the table.

Rectangle	Area
A	
B	
C	

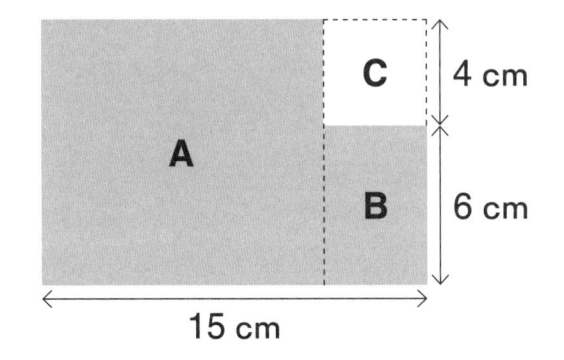

(b) Find the area of the remaining piece of paper by adding areas A and B.

(c) Find the area of the paper before it was cut.

(d) Find the area of the remaining piece of paper by subtracting area C from the area of the paper before it was cut.

(e) A 4-cm square is cut from the other two pieces of paper as shown. What is the area of each remaining piece of paper?

11-3 Area of Composite Figures

2 Find the unknown lengths. Then find the area of the shaded part of the rectangle.

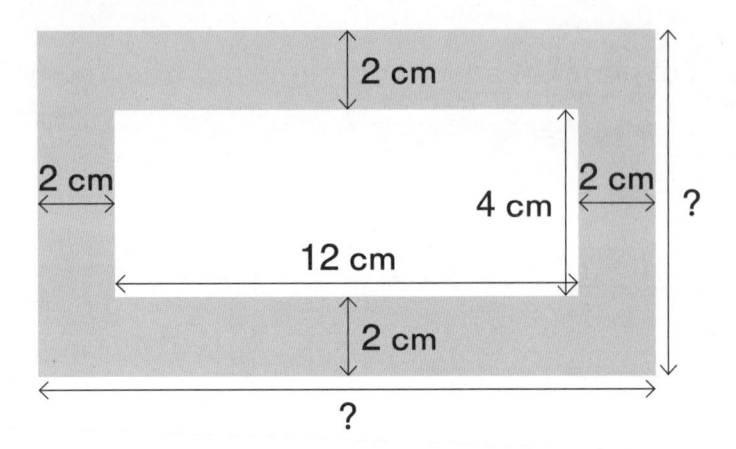

Practice

3 Find the area of each figure.

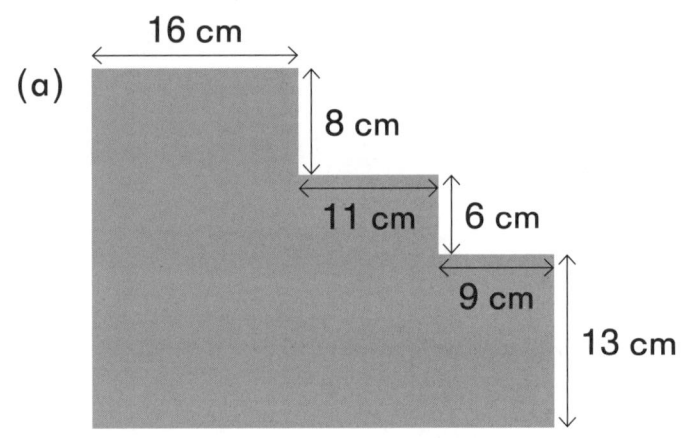

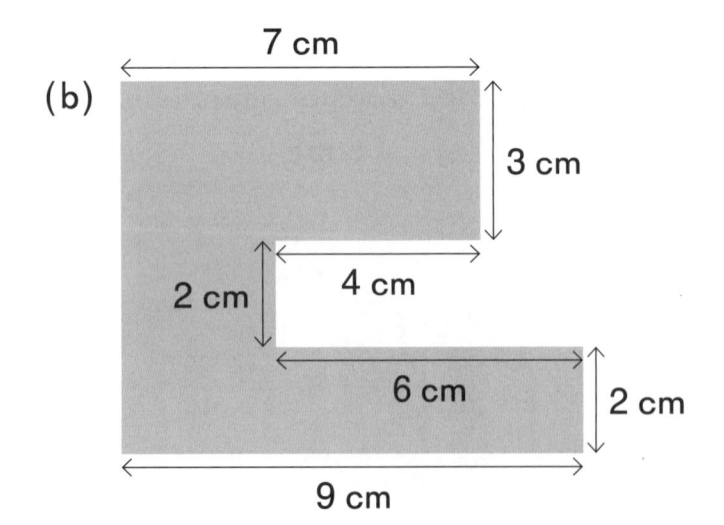

(a)

(b)

4 What is the area of the shaded part of the figure in square inches?

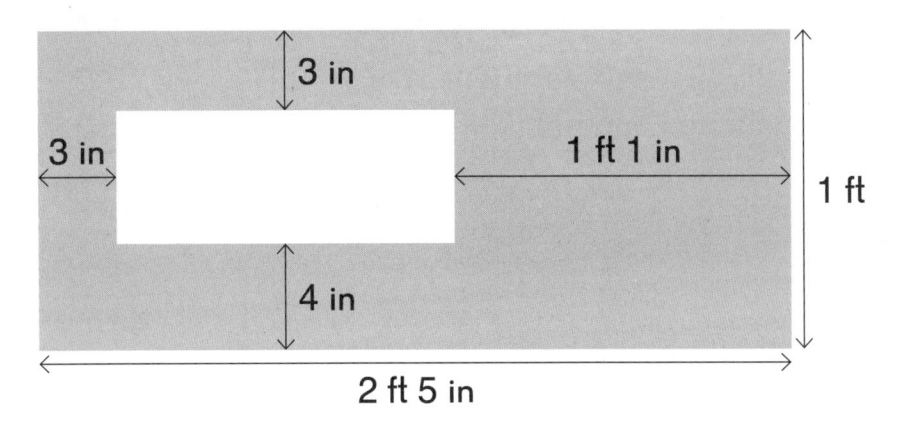

- 3 in (top)
- 3 in (left)
- 1 ft 1 in
- 1 ft (right)
- 4 in (bottom)
- 2 ft 5 in (total width)

5 A rectangular field is 13 m long and 10 m wide. It has a cement path $3\frac{1}{2}$ m wide around it. What is the area of the cement path in square meters?

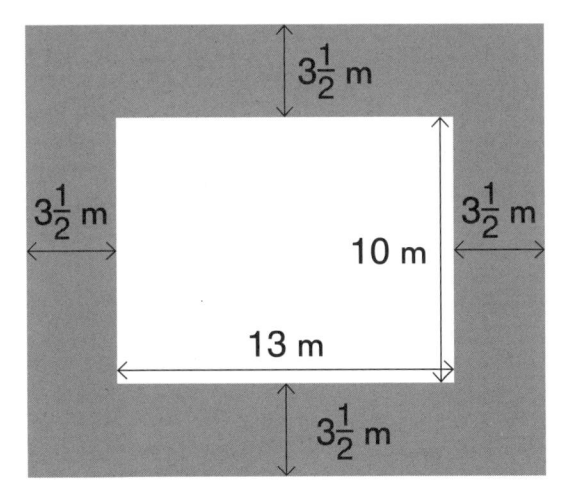

- $3\frac{1}{2}$ m (top)
- $3\frac{1}{2}$ m (left)
- $3\frac{1}{2}$ m (right)
- $3\frac{1}{2}$ m (bottom)
- 10 m
- 13 m

Challenge

6 The figure is made up of five overlapping squares, each with an area of 80 cm². The total area of the shaded parts is 360 cm². Find the area of each of the smaller, unshaded squares.

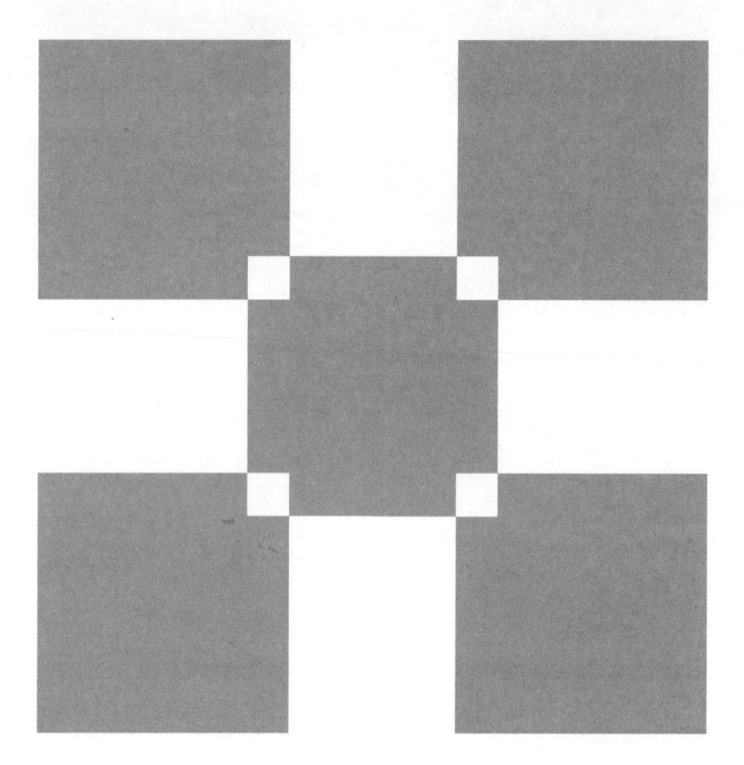

Exercise 4

Basics

1 The perimeter of a rectangle is 40 cm. One side is 7 cm long. What is the length of the other side?

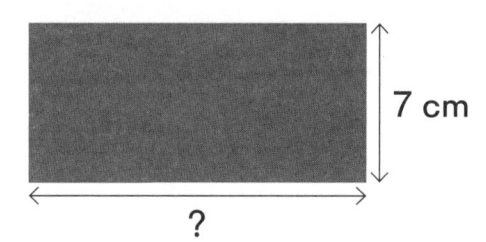

7 cm

?

2 × Width = 2 × 7 = ☐ cm

2 × Length = 40 − ☐ = ☐ cm

Length = ☐ ÷ 2 = ☐ cm

2 The perimeter of a rectangle is 160 cm. One side is 45 cm long. How long is the other side?

?

45 cm

Length + Width = 160 ÷ 2 = ☐ cm

Width = ☐ − 45 = ☐ cm

3 A square has a perimeter of 136 cm. What is the length of one side of the square?

?

Side = 136 ÷ 4 = ☐ cm

11-4 Perimeter — Part 1   39

Practice

4 Complete the table.

Rectangle	Area	Length	Width	Perimeter
A		15 cm	12 cm	
B	120 cm²		10 cm	
C		10 cm		120 cm
D	198 cm²		9 cm	
E			45 cm	198 cm

5 The perimeter of a rectangle is 8 ft 8 in. Its width is 1 ft 9 in. Express its length in feet and inches.

1 ft 9 in

?

6 Evan ran around a rectangular field twice. He ran a total distance of 1 km 620 m. The width of the field is 195 m. What is the length of the field?

7) A square 1 ft long and a rectangle 8 in wide have the same perimeter. What is the length of the rectangle?

Challenge

8) The figure is made up of 3 identical rectangles. The area of the whole figure is 216 cm². What is the perimeter of the whole figure?

Exercise 5

Basics

1 (a) Find the perimeter of each figure by adding all the side lengths.

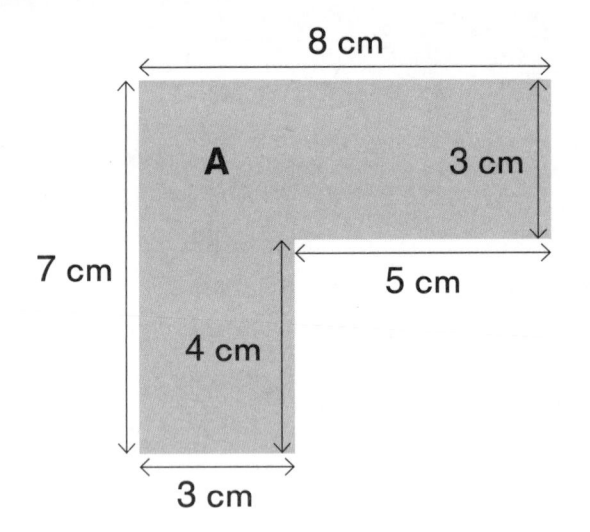

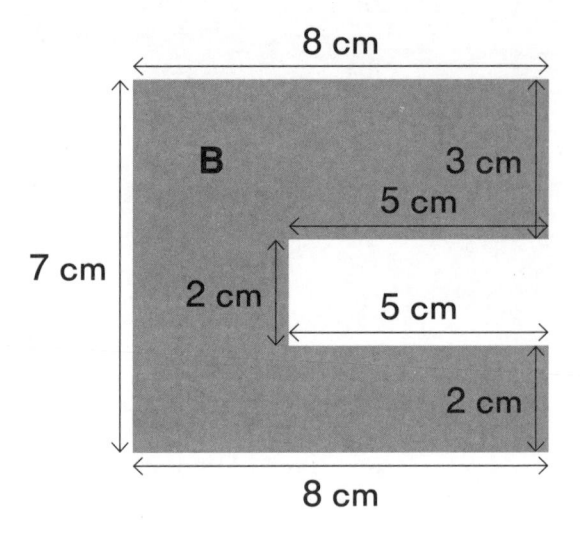

Figure A: 8 cm (top), 3 cm (right), 5 cm, 4 cm, 3 cm (bottom), 7 cm (left)

Figure B: 8 cm (top), 3 cm, 5 cm, 2 cm, 5 cm, 2 cm, 8 cm (bottom), 7 cm (left)

(b) Compare the perimeters of Figures A and B to Rectangle C. What do you notice? Use this knowledge to find the perimeter of Figure D.

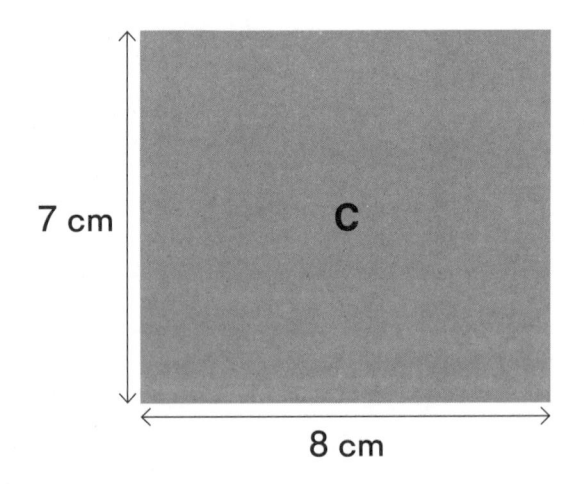

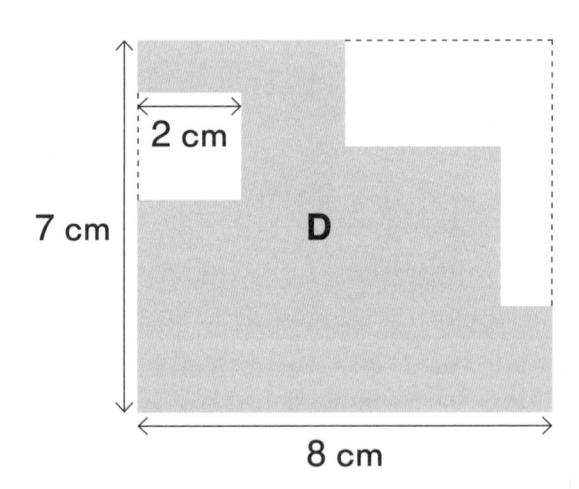

Rectangle C: 7 cm by 8 cm

Figure D: 7 cm (left), 8 cm (bottom), with 2 cm notch

42   11-5 Perimeter — Part 2

Practice

2 The sides of the figures are marked in units. Find all the sets of figures that have equal perimeters.

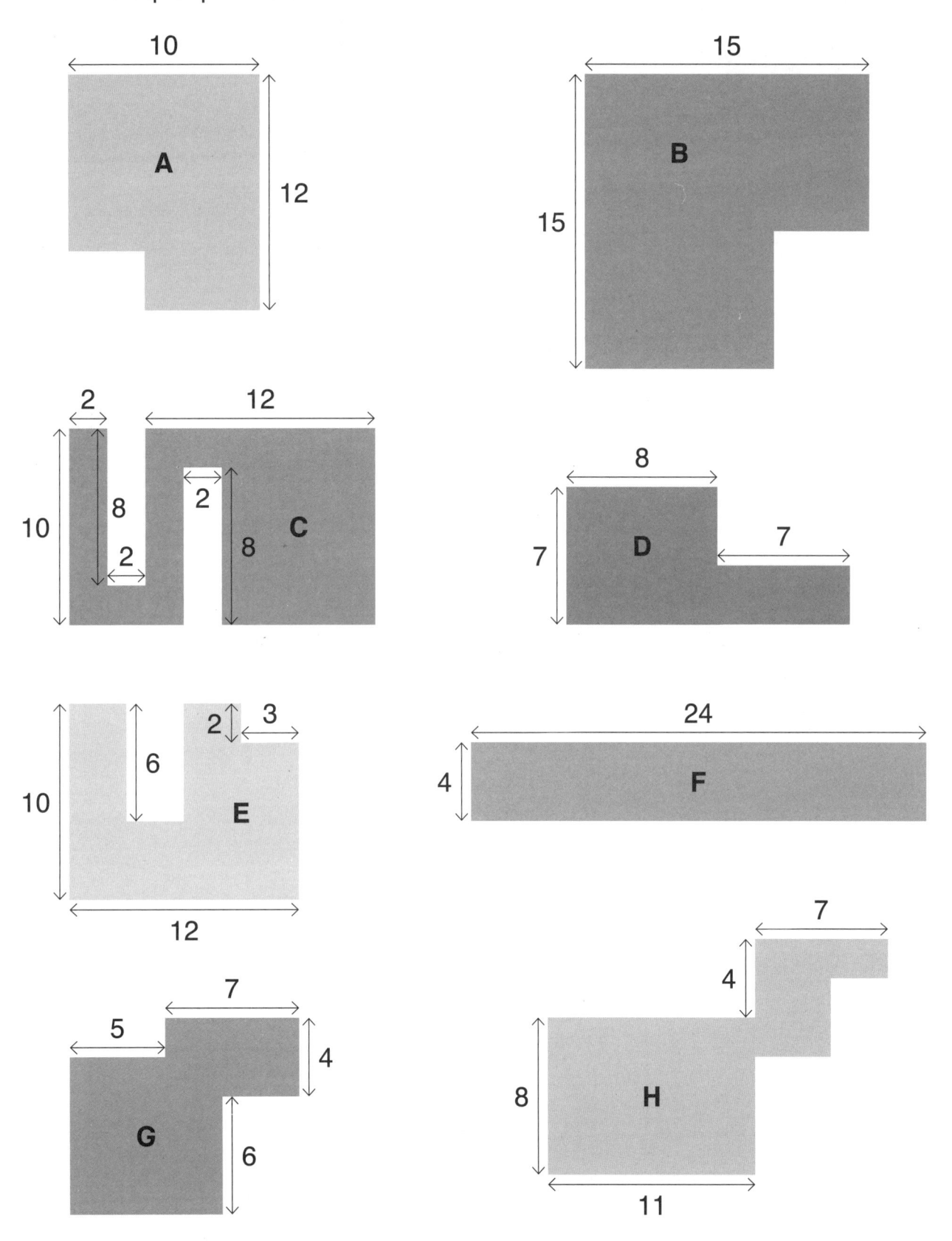

3. Each step of this staircase has the same length and height as the first step. What is the perimeter of the figure?

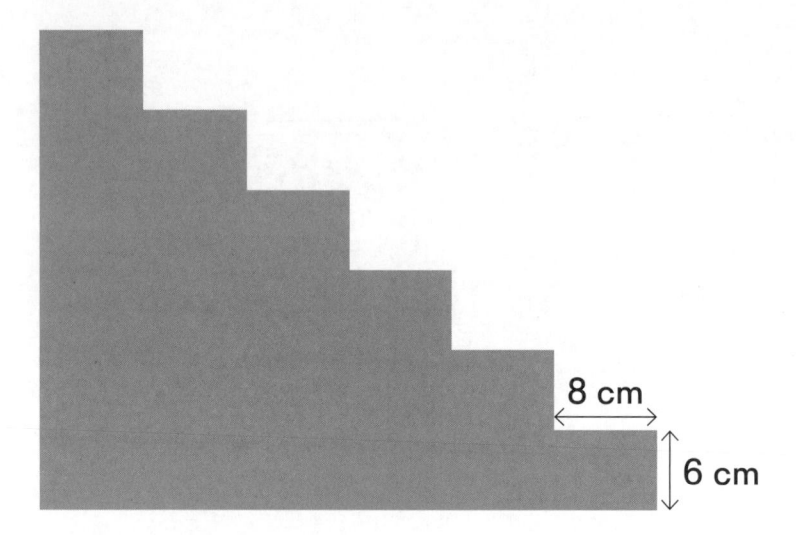

8 cm
6 cm

Challenge

4. Each rectangle is 4 in long and has an area of 8 in². What is the perimeter of the figure?

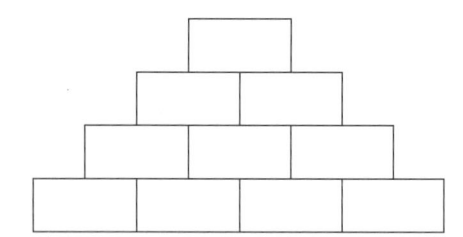

Exercise 6

Check

1 It costs $12 per square foot to carpet a room. The perimeter of the room is 54 ft and one side of the room is 15 ft. How much will it cost to carpet the room?

2 A rectangular beach towel is 1 m 10 cm long and 90 cm wide. It has a border that is 3 cm wide around it. What is the area of the border?

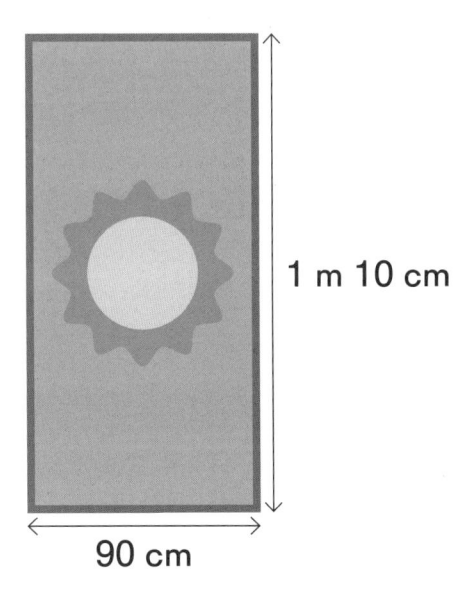

1 m 10 cm

90 cm

3 Amanda has a poster board that is 3 ft long by 2 ft wide. She cuts out 4 identical squares from each corner with sides measuring $\frac{1}{2}$ ft. Find the area of the remaining piece of poster board in square inches and the perimeter in inches.

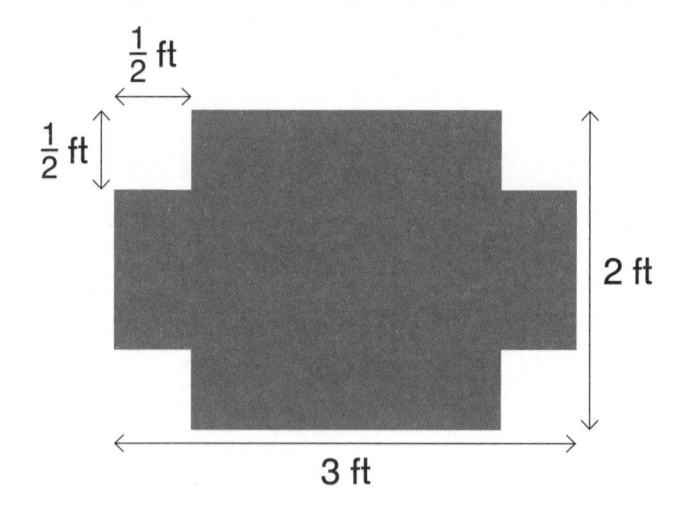

4 Jamal has a poster board that is 36 cm by 20 cm. He cuts out 3 squares from one side. The largest square is 8 cm long. Find the area and perimeter of the remaining piece of poster board.

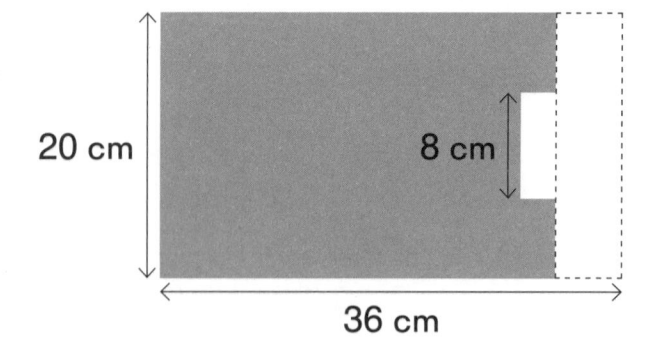

5 Ivy has a poster board 3 ft long and 2 ft wide. She glues a rectangular picture in the middle that is 3 inches from each width and 4 inches from each length. What is the area of the picture?

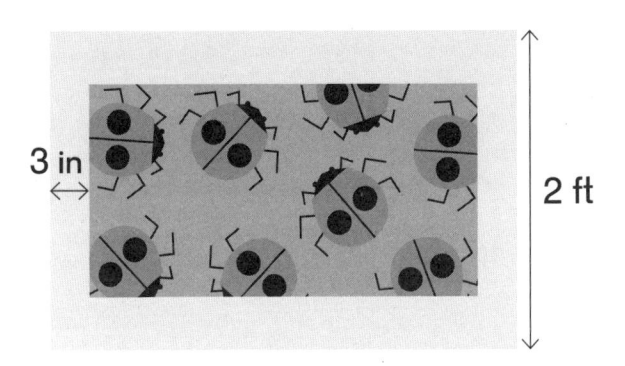

3 in

2 ft

6 This figure is made of triangles with all three sides of equal length. The perimeter of each triangle is 2 ft. What is the perimeter of the figure? Express the answer in feet and inches.

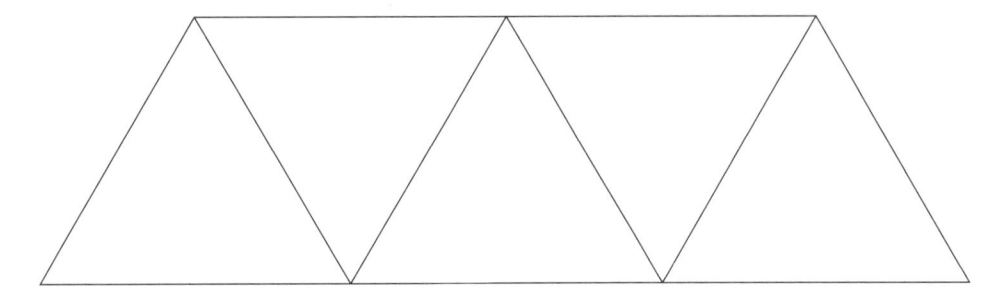

Challenge

7 Aaron glued 4 identical strips of paper to form a square. The perimeter of each rectangular strip of paper is 20 cm. What is the area of the larger square formed?

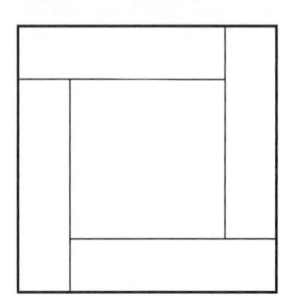

8 A 40 cm by 50 cm rectangular flag is painted with 3 stripes, each 5 cm wide. What is the area of the parts that are not painted?

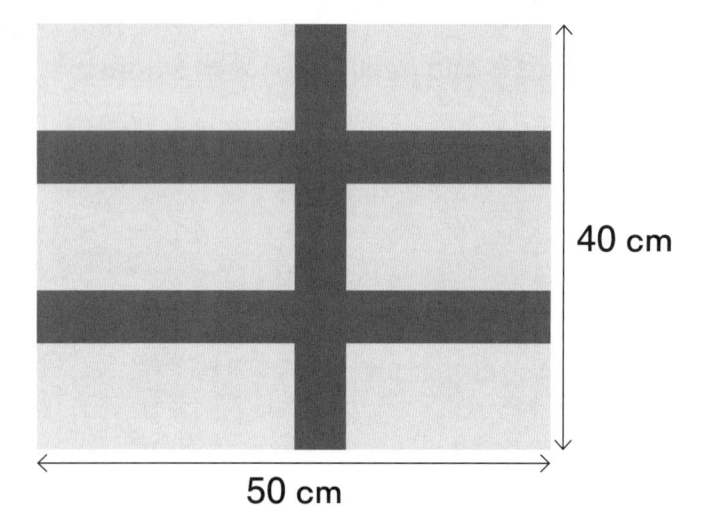

40 cm

50 cm

Chapter 12 Decimals

Exercise 1

Basics

1 Complete the table.

	Fraction	Decimal
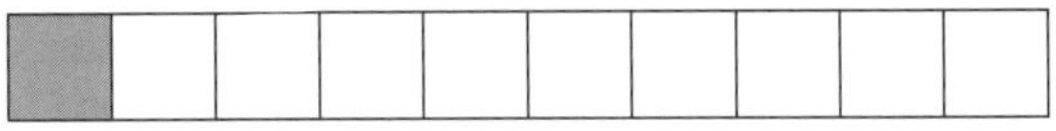	$\frac{1}{10}$	0.1
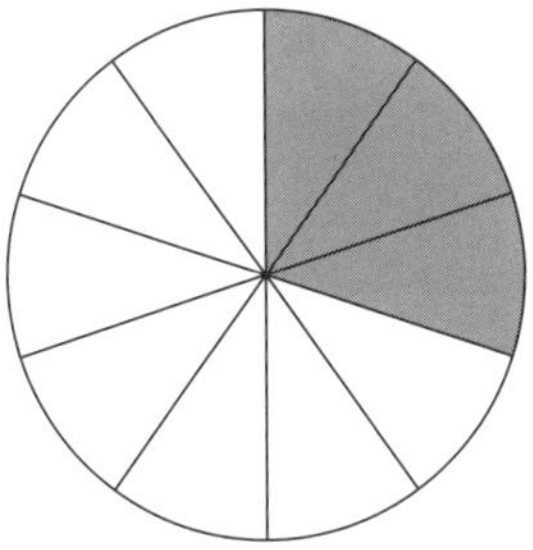		
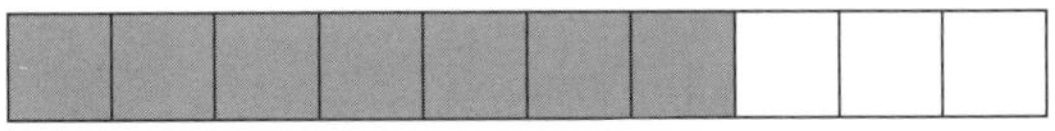		
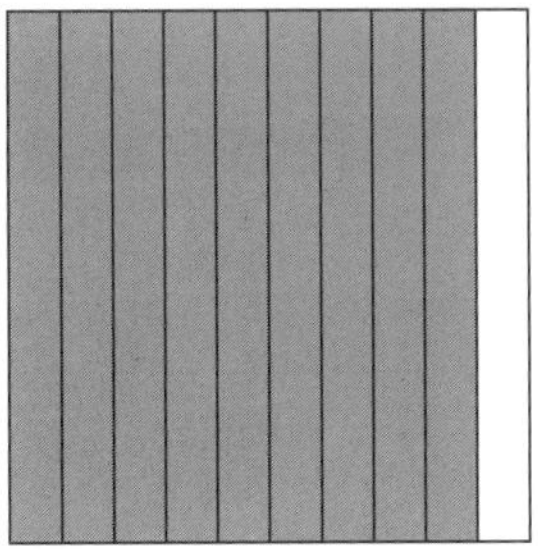		

2 Write a decimal to make the equations true.

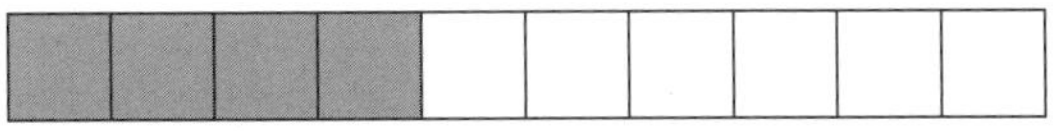

(a) 0.4 + ☐ = 1

(b) 1 – 0.4 = ☐

Practice

3 Shade each square to show the decimal.

(a) 0.5

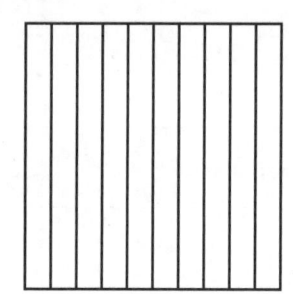

(b) 0.8

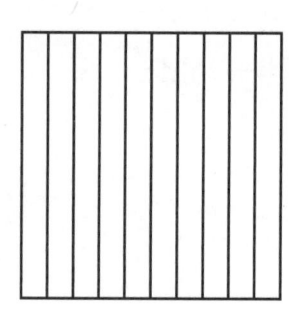

4 Write the decimal indicated by each arrow.

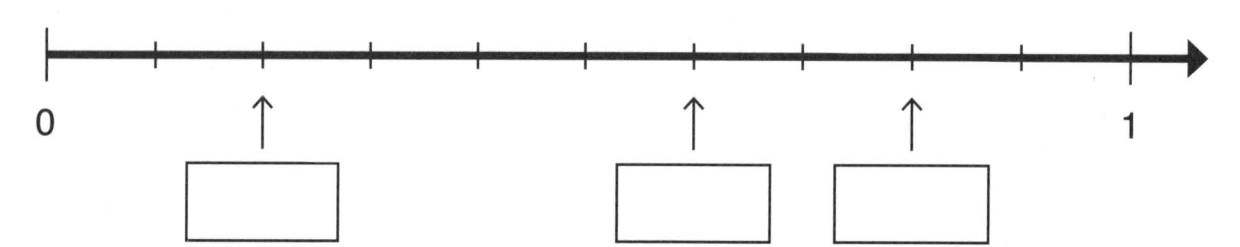

5 Write the decimals represented by the place-value discs.

(a)

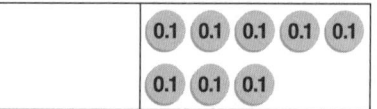

(b)

Ones	Tenths

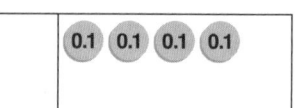

Ones	Tenths

6 Write each fraction as a decimal.

(a) $\frac{3}{10} =$

(b) $\frac{1}{10} =$

(c) $\frac{9}{10} =$

(d) $\frac{40}{100} =$

7 Max ran 0.7 km. He wants to run 1 km. How much farther does he have to run?

8 (a) ☐ mm = 1 cm

(b) 1 mm = ☐/☐ cm

(c) 1 mm = ☐ cm

(d) 4 mm = ☐ cm

Challenge

9 1 deciliter (dL) is $\frac{1}{10}$ of 1 liter. Pierre has a bottle with 1 L of water. He drank 2 deciliters of water.

(a) Express the amount he drank as a fraction of a liter.

(b) Express the amount he drank in liters as a decimal.

(c) Express the amount of water left in liters as a decimal.

Exercise 2

Basics

1

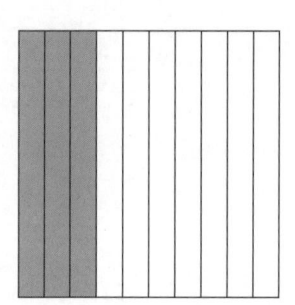

2 ones + 3 tenths = 2 + $\dfrac{\Box}{10}$ = 2$\dfrac{\Box}{}$

2 ones + 3 tenths = 2 + 0.3 = 2.$\Box$

2

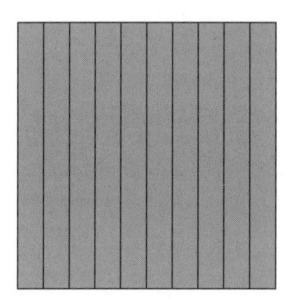

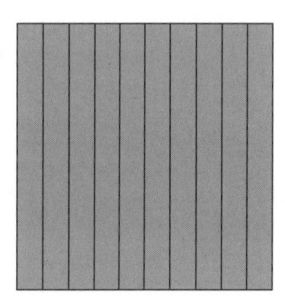

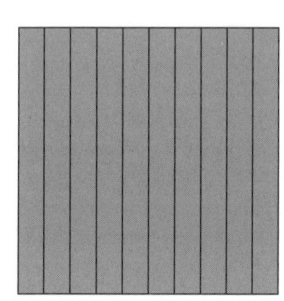

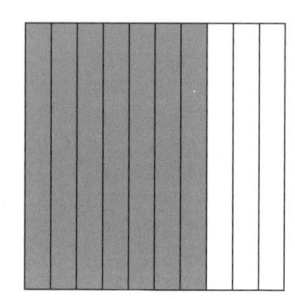

37 tenths = $\dfrac{\Box}{10}$ = 3$\dfrac{\Box}{}$

37 tenths = $\Box$ ones $\Box$ tenths = 3 + 0.$\Box$ = 3.$\Box$

3 Write each measurement as a decimal.

(a)

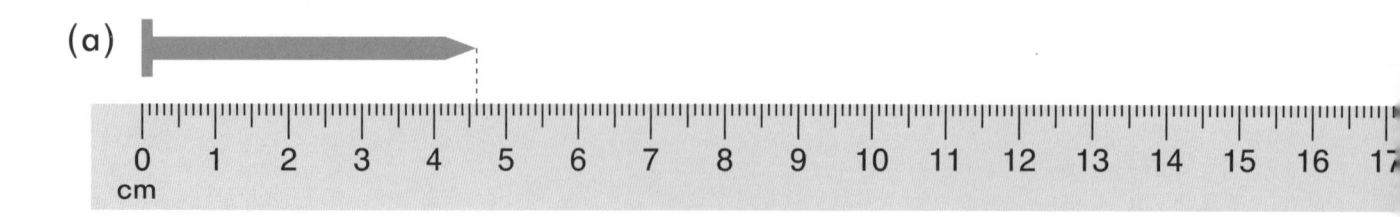

The length of the nail is _____ cm.

(b)

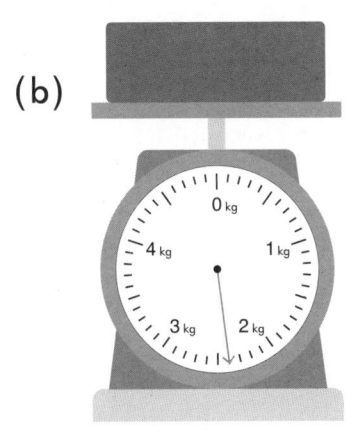

The block weighs _____ kg.

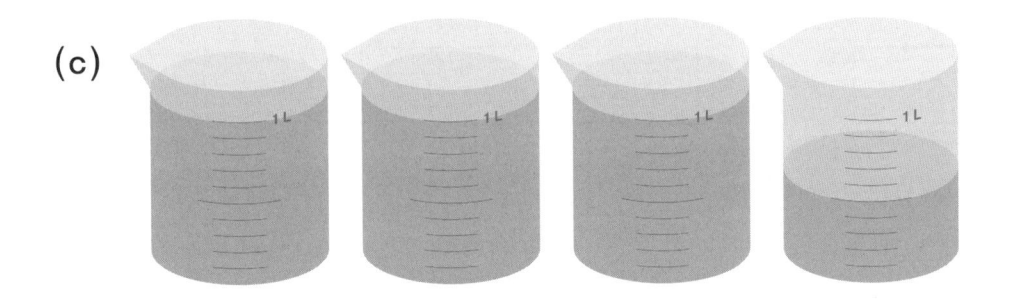

(c)

There is a total of _____ L of water.

4. Write the decimals represented by the place-value discs.

(a)

Tens	Ones	Tenths
		•

(b)

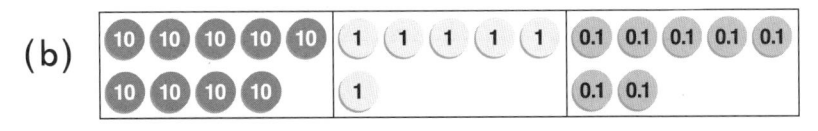

Tens	Ones	Tenths
		•

Practice

5 Write the decimals indicated by each arrow.

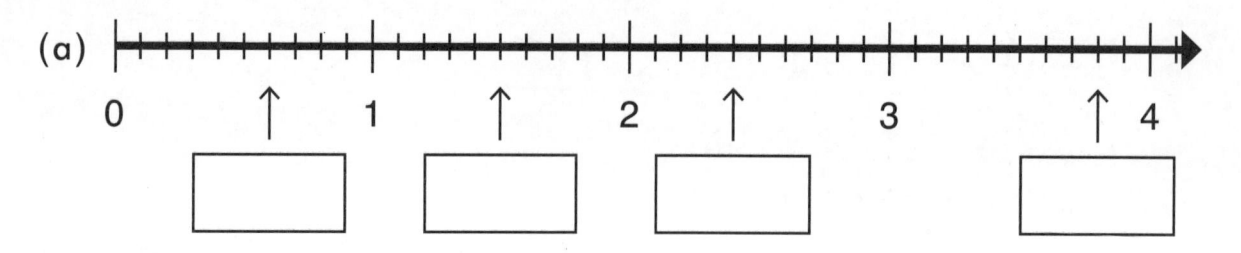

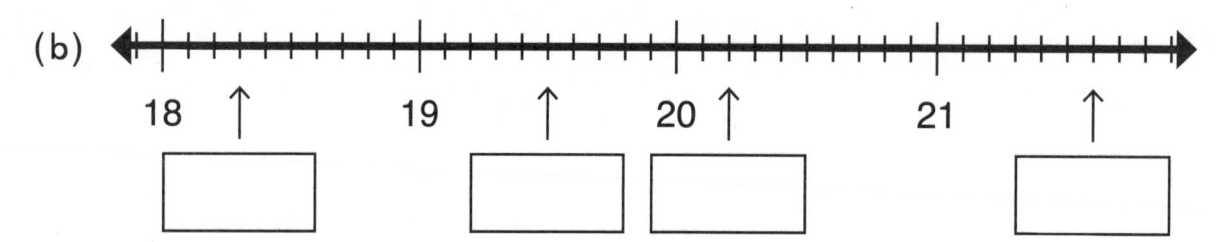

(a) Number line from 0 to 4 with arrows pointing to positions between whole numbers.

(b) Number line from 18 to 21+ with arrows pointing to positions between whole numbers.

6 Write each number indicated by the place-value discs as a decimal and as a mixed number (or whole number).

	Decimal	Mixed Number
10 discs of 0.1 and 9 discs of 0.1 (19 discs of 0.1)		
9 discs of 1 and 1 disc of 0.1		
10 discs of 0.1 and 11 discs of 0.1 (21 discs of 0.1)		
5 discs of 10 and 7 discs of 0.1 and 2 discs of 0.1		

54   12-2 Tenths — Part 2

7 Complete the number patterns.

(a) | 3.7 | 3.8 | 3.9 | | | |

(b) | 10.2 | 10.1 | | | 9.8 | |

(c) | 38.2 | | | 41.2 | 42.2 | |

8 (a) 9 + 0.4 = ☐ (b) 60 + 1 + 0.8 = ☐

(c) 0.5 + 30 = ☐ (d) 900 + 8 + 0.1 = ☐

(e) 20 + 9 + ☐ = 29.7 (f) 0.5 + ☐ = 4.5

Challenge

9 Write >, <, or = in each ◯.

(a) $4\frac{3}{10}$ ◯ 3.4 (b) 70 ◯ 7.0

(c) 21 tenths ◯ 2.1 (d) 6.7 ◯ $\frac{60}{10}$

(e) 6 ◯ 0.6 (f) 1.3 ◯ 130 tenths

10 (a) 12.3 = ☐ tenths (b) 132.4 = ☐ tenths

12-2 Tenths — Part 2   55

Exercise 3

Basics

1 Complete the table.

	Fraction	Decimal
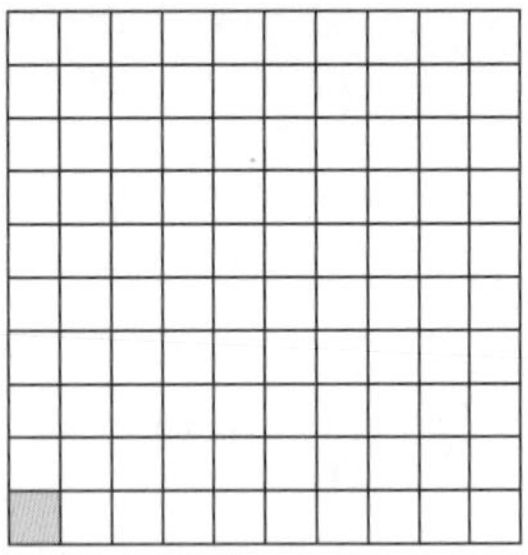	$\frac{1}{100}$	0.01
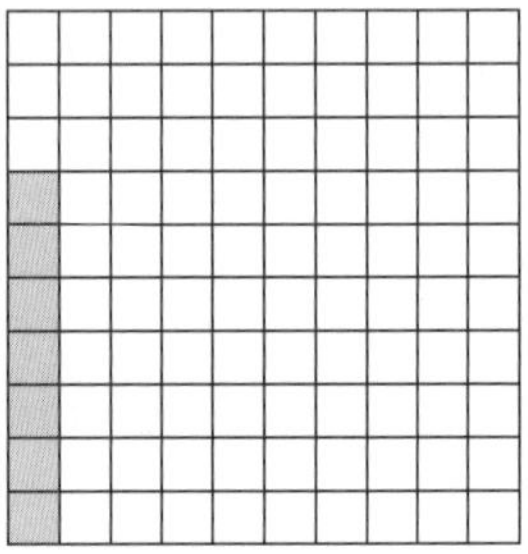	$\frac{\square}{100}$	$\square$
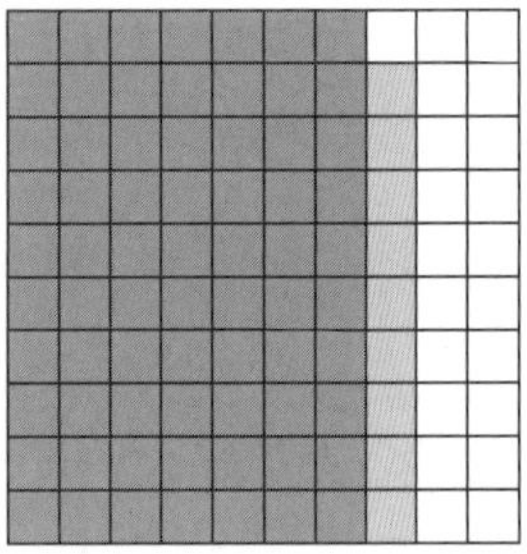	$\frac{\square}{10} + \frac{\square}{100}$	$\square$
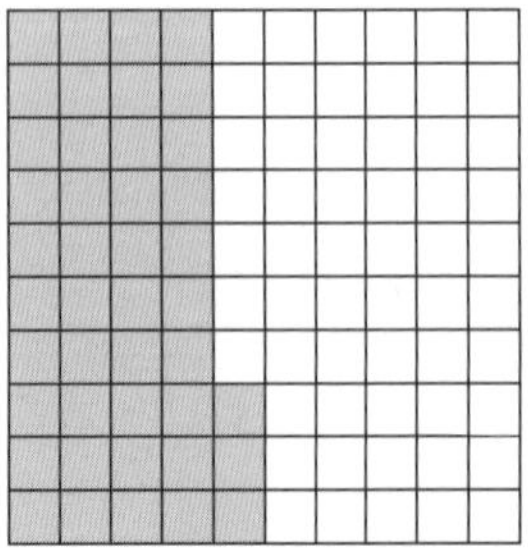	$\frac{\square}{100}$	$\square$

2. Shade each square to show the decimal.

(a) 0.28

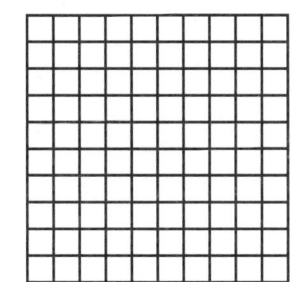

(b) 0.92

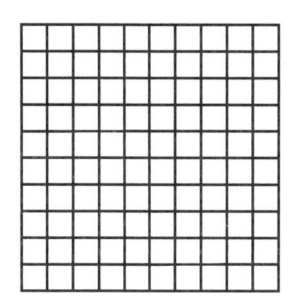

Practice

3. Write the decimals indicated by each arrow.

0.4   0.5   0.6   0.7   0.8

4. Write the decimals represented by the place-value discs.

(a)

Ones	Tenths	Hundredths

(b)

Ones	Tenths	Hundredths

12-3 Hundredths — Part 1 57

5 Complete the table.

	Decimal	Fraction
(17 × 0.01)		
(5 × 0.1) + (12 × 0.01)		

6 Complete the number patterns.

(a) | 0.17 | 0.18 | 0.19 | | | |

(b) | 0.64 | 0.63 | 0.62 | | | |

7 Write the decimal.

(a) $\frac{6}{100}$ =

(b) $\frac{56}{100}$ =

(c) $\frac{8}{10} + \frac{7}{100}$ =

(d) 0.6 + 0.08 =

(e) 1 − 0.65 =

(f) 0.51 + ⬚ = 1

8 Express each length as a decimal.

(a) 4 cm = ⬚ m

(b) 99 cm = ⬚ m

58   12-3 Hundredths — Part 1

Exercise 4

Basics

1 Complete the table.

	Fraction	Decimal
	$1 + \dfrac{\boxed{}}{10} + \dfrac{\boxed{}}{100}$	1.36
	$2 + \dfrac{\boxed{}}{100}$	
	$2\dfrac{\boxed{}}{100}$	
	$1 + \dfrac{\boxed{}}{10} + \dfrac{\boxed{}}{100}$	
	$\dfrac{\boxed{}}{100}$	

12-4 Hundredths — Part 2

2 Shade the squares to show the decimal.

(a) 1.28

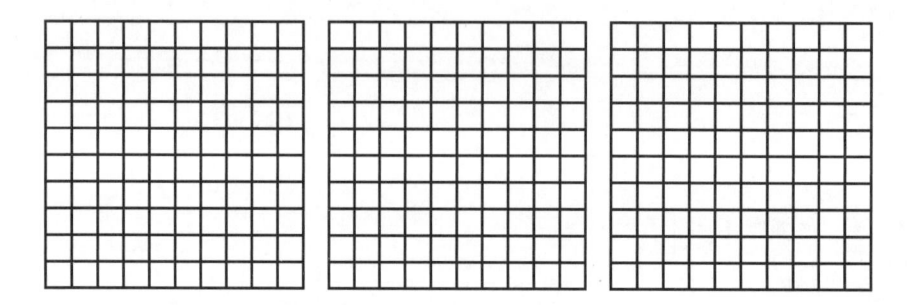

(b) 2.71

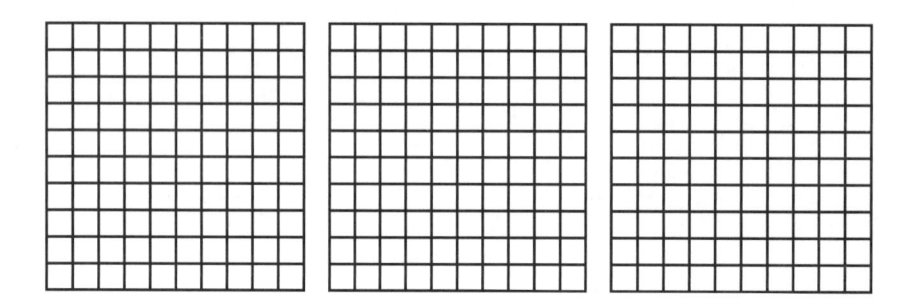

3 (a) Write the decimal represented by the place-value discs.

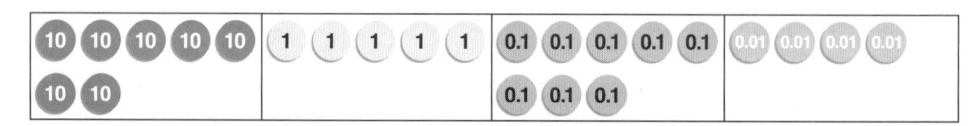

Tens	Ones	Tenths	Hundredths

(b) The digit _____ is in the tens place. Its value is _____.

(c) The digit _____ is in the ones place. Its value is _____.

(d) The digit _____ is in the _____ place. Its value is 0.8.

(e) The digit 4 is in the _____ place. Its value is _____.

(f) 70 + 5 + 0.8 + 0.04 = ☐

Practice

4 Write the decimal indicated by each arrow.

(a)

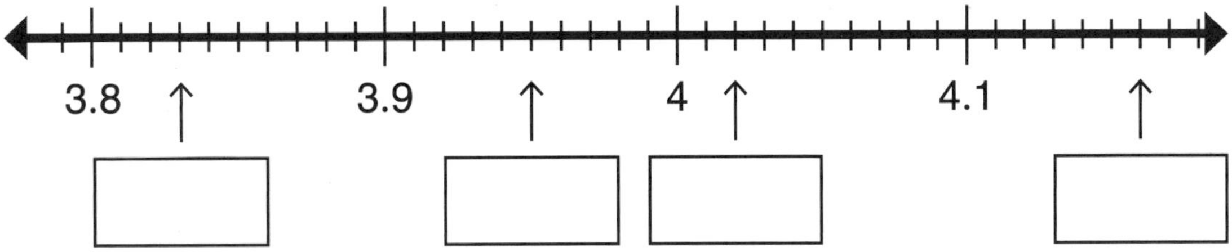

(b)

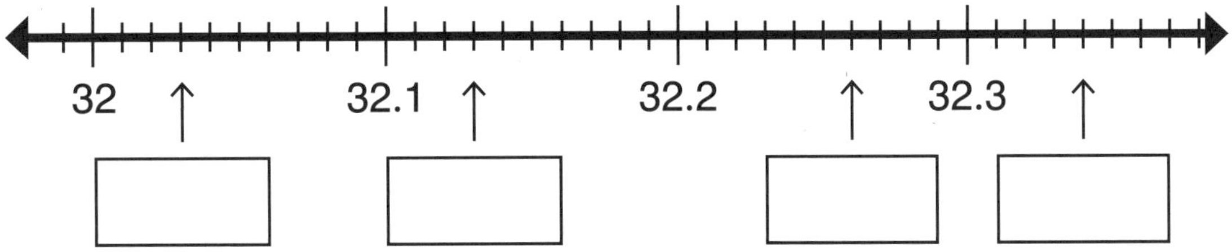

5 Write the decimals represented by the place-value discs. Then write each number in expanded form.

(a)

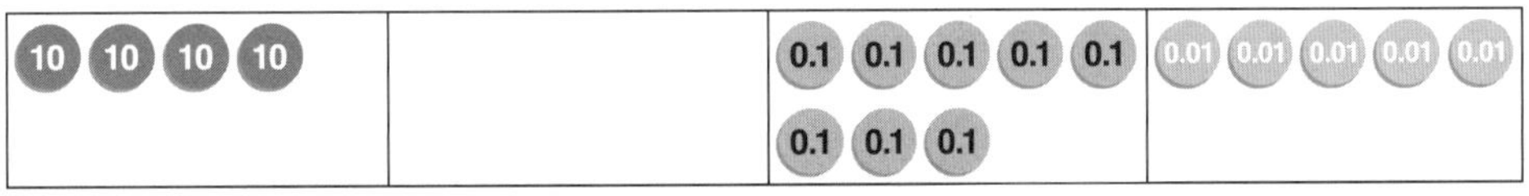

Tens	Ones	Tenths	Hundredths
	•		

Expanded form:

(b)

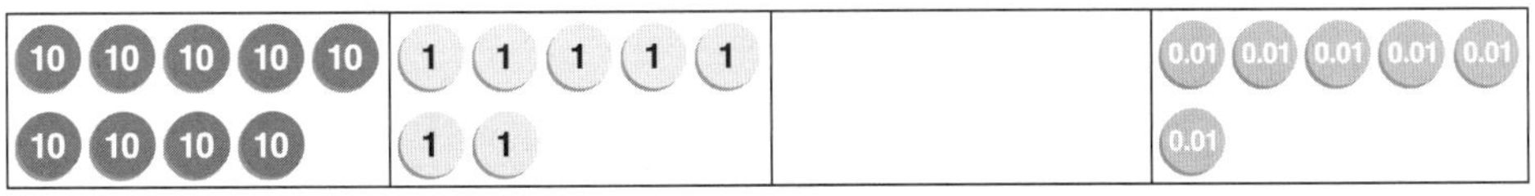

Tens	Ones	Tenths	Hundredths
	•		

Expanded form:

6 Complete the number patterns.

(a) | 4.97 | 4.98 | 4.99 | ⬚ | ⬚ | ⬚ |

(b) | 8.62 | 8.61 | ⬚ | ⬚ | ⬚ | 8.57 |

(c) | 29.62 | 29.72 | ⬚ | 29.92 | ⬚ | ⬚ |

7 Write the decimal.

(a) $7\frac{9}{100}$ = ⬚

(b) $3\frac{56}{100}$ = ⬚

(c) $\frac{170}{100}$ = ⬚

(d) 7 + 0.6 + 0.08 = ⬚

(e) 500 + 0.8 + 0.04 = ⬚

(f) 30 + 7 + ⬚ + 0.01 = 37.31

(g) ⬚ + 6 + 40 = 46.65

(h) six and four hundredths = ⬚

Exercise 5

Basics

1 Express 0.8 as a fraction in simplest form.

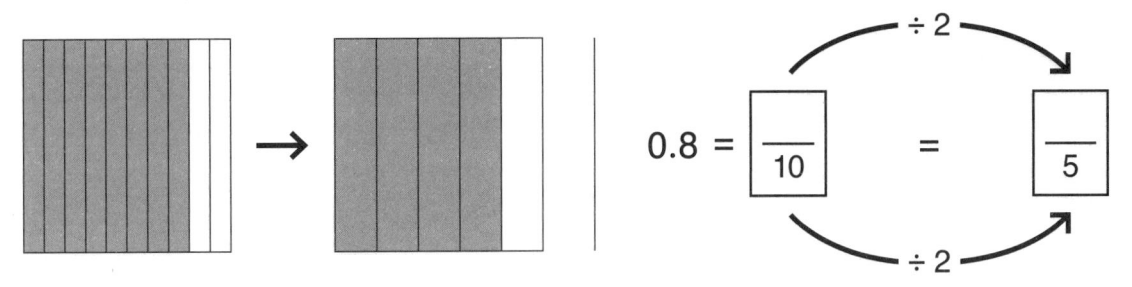

$$0.8 = \frac{}{10} = \frac{}{5}$$

(÷ 2)

2 Express 0.45 as a fraction in simplest form.

$$0.45 = \frac{}{100} = \frac{}{}$$

(÷ 5)

3 Express 0.32 as a fraction in simplest form.

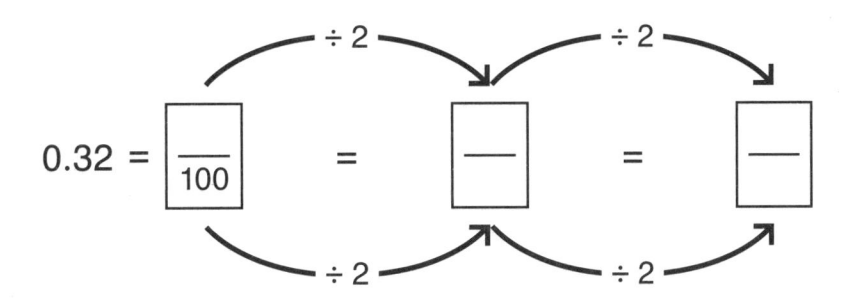

$$0.32 = \frac{}{100} = \frac{}{} = \frac{}{}$$

(÷ 2, ÷ 2)

4 Express 4.48 as a mixed number in simplest form.

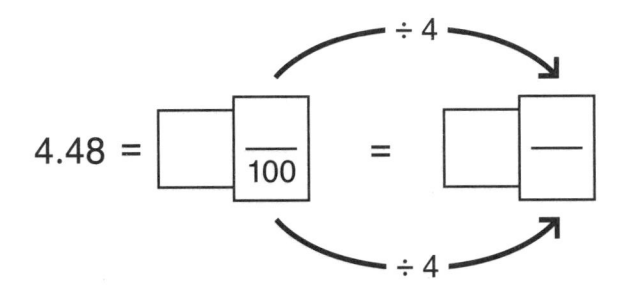

$$4.48 = \boxed{}\frac{}{100} = \boxed{}\frac{}{}$$

(÷ 4)

12-5 Expressing Decimals as Fractions in Simplest Form

Practice

5 Finish labeling each tick mark with a decimal and with a fraction in simplest form.

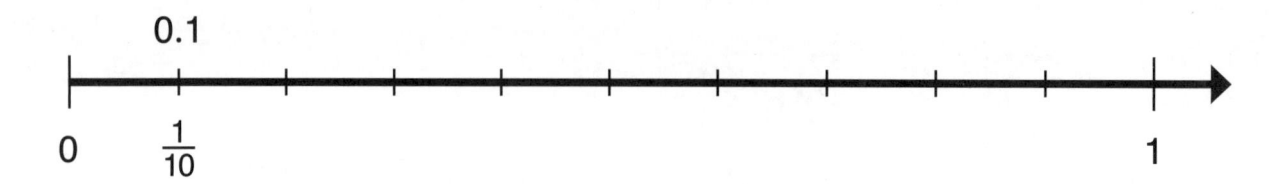

```
        0.1
|---+---+---+---+---+---+---+---+---+---+---→
0    1/10                                    1
```

6 Match.

4.16	$4\frac{17}{20}$
4.62	$4\frac{18}{25}$
4.05	$4\frac{4}{25}$
4.85	$4\frac{31}{50}$
4.20	$4\frac{1}{20}$
4.72	$4\frac{1}{5}$

7 Write each decimal as a fraction or mixed number in simplest form.

(a) 6.4

(b) 0.85

(c) 8.08

(d) 10.54

8 A beaker contains 2.35 L of solution. Express this amount in liters as a mixed number in simplest form.

9 A package weighs 7.64 kg. Express this weight in kilograms as a mixed number in simplest form.

10 A football is 10.5 in long and the white stripe is 2.25 in from the end of the ball. Express each measurement in inches as a mixed number in simplest form.

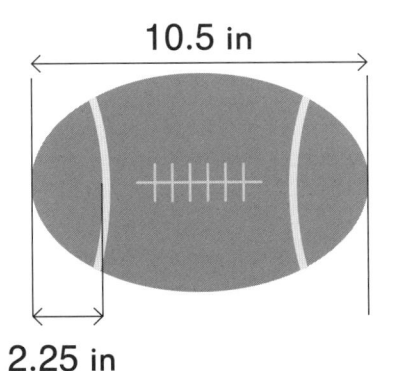

10.5 in

2.25 in

Exercise 6

Basics

1 Express $\frac{1}{2}$ as a decimal.

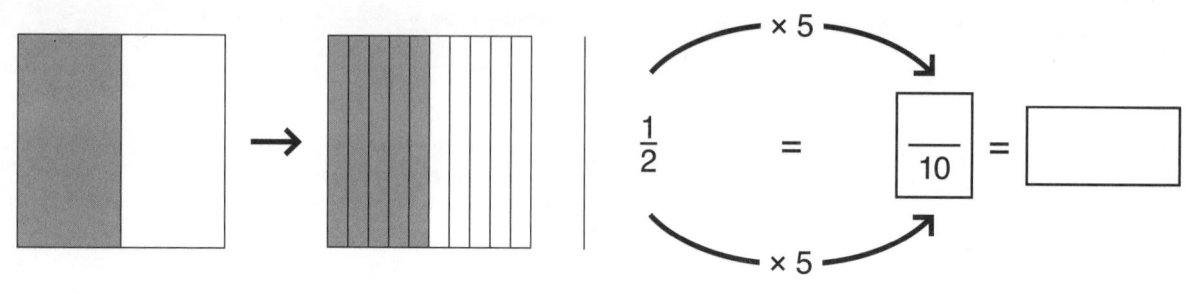

$\frac{1}{2} = \frac{\boxed{}}{10} = \boxed{}$ (×5 / ×5)

2 Express $\frac{17}{20}$ as a decimal.

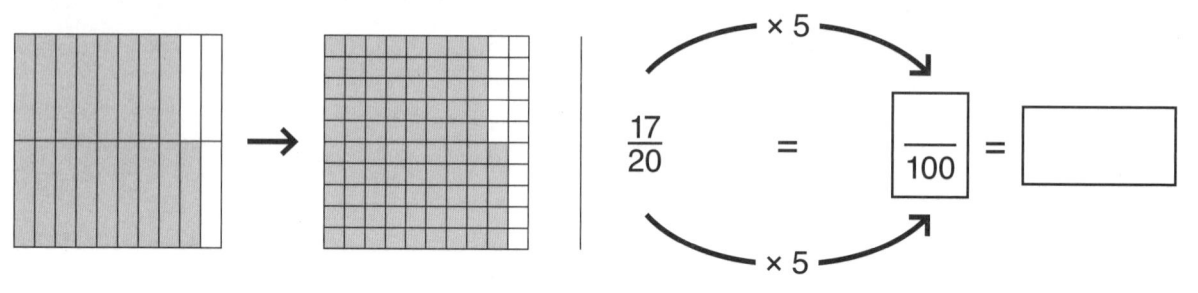

$\frac{17}{20} = \frac{\boxed{}}{100} = \boxed{}$ (×5 / ×5)

3 Express $\frac{9}{5}$ as a decimal.

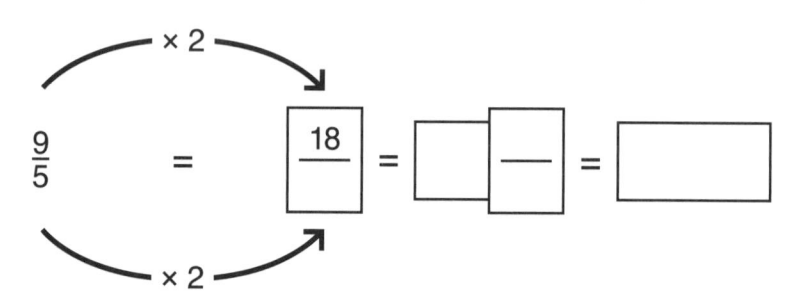

$\frac{9}{5} = \frac{18}{\boxed{}} = \boxed{}\frac{\boxed{}}{\boxed{}} = \boxed{}$ (×2 / ×2)

4 Express $7\frac{6}{25}$ as a decimal.

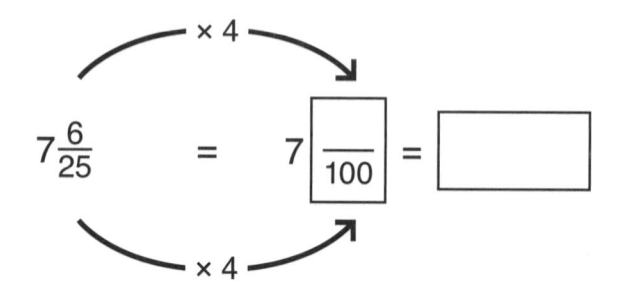

$7\frac{6}{25} = 7\frac{\boxed{}}{100} = \boxed{}$ (×4 / ×4)

12-6 Expressing Fractions as Decimals

Practice

5 Express each fraction as a decimal.

$\frac{1}{2}$ = ☐

$\frac{1}{4}$ = ☐ $\frac{3}{4}$ = ☐

$\frac{1}{5}$ = ☐ $\frac{2}{5}$ = ☐ $\frac{3}{5}$ = ☐ $\frac{4}{5}$ = ☐

$\frac{1}{20}$ = ☐ $\frac{7}{20}$ = ☐ $\frac{13}{20}$ = ☐ $\frac{19}{20}$ = ☐

$\frac{1}{25}$ = ☐ $\frac{7}{25}$ = ☐ $\frac{13}{25}$ = ☐ $\frac{24}{25}$ = ☐

$\frac{1}{50}$ = ☐ $\frac{7}{50}$ = ☐ $\frac{13}{50}$ = ☐ $\frac{49}{50}$ = ☐

6 A football weighs $14\frac{2}{5}$ oz. Express the weight of the football as a decimal.

7 The wingspan of a house swallow is $8\frac{3}{4}$ in. Express this length as a decimal.

12-6 Expressing Fractions as Decimals   67

Exercise 7

Check

1 In the number 98.95:

(a) The value of the digit in the tenths place is _____.

(b) The value of the digit in the tens place is _____ times the value of the digit in the tenths place.

(c) The digit with a value of 0.05 is in the _____ place.

(d) Write the number in expanded form.

2 (a) 8 + ☐ + 0.06 = 8.26

(b) ☐ + 0.6 + 72 = 72.64

3 Express each value as a fraction or as a mixed number in simplest form.

(a) 0.3

(b) 0.07

(c) 0.21

(d) 4.79

(e) 8.90

(f) 6.5

(g) 6.15

(h) 11.04

4 Express each value as a decimal.

(a) $\frac{4}{10}$

(b) $10\frac{7}{10}$

(c) $\frac{25}{100}$

(d) $7\frac{52}{100}$

(e) $\frac{37}{10}$

(f) $\frac{620}{100}$

(g) $2\frac{3}{4}$

(h) $6\frac{17}{20}$

(i) $\frac{7}{2}$

(j) $\frac{32}{25}$

5 (a) What number is 2 hundredths more than 3.99?

(b) What number is 2 tenths more than 3.99?

(c) What number is 2 hundredths less than 6.01?

(d) What number is 2 tenths less than 6.01?

Challenge

6 Express each value as a decimal.

(a) $3 + \frac{2}{5} + \frac{2}{25}$

(b) $8 + \frac{1}{2} + \frac{3}{50}$

Exercise 8

Basics

1 Shade the squares to show each decimal. Then write > or < in each ◯.

(a) 0.35 ◯ 0.53

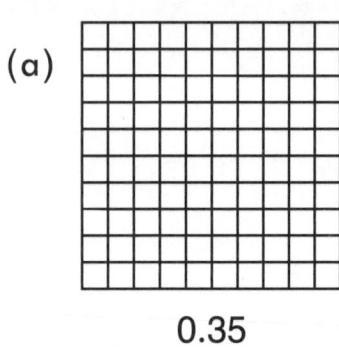

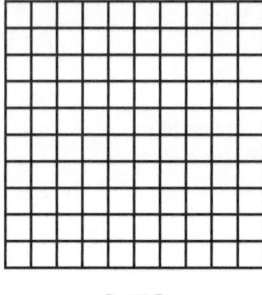

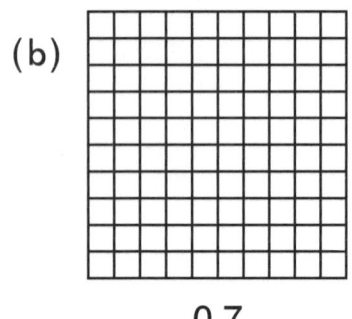

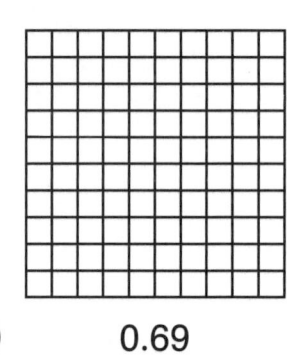

(b) 0.7 ◯ 0.69

2 Circle the digits in the greatest place that are different in each number. Then write > or < in each ◯.

(a)

Ones	Tenths	Hundredths
1	5	2
1	8	1

0.5 ◯ 0.8

1.52 ◯ 1.81

(b)

Ones	Tenths	Hundredths
4	6	7
4	6	0

0.07 ◯ 0

4.67 ◯ 4.6

3 Write any missing digits, then write > or < in each ◯.

(a) Compare $3\frac{4}{5}$ and 3.45

$3\frac{4}{5} = 3.\boxed{}$ | $3.\boxed{} \bigcirc 3.45$ | $3\frac{4}{5} \bigcirc 3.45$

(b) Compare $2\frac{1}{8}$ and 2.75

$2.75 = 2\boxed{\frac{}{}}$ | $2\frac{1}{8} \bigcirc 2\boxed{\frac{}{}}$ | $2\frac{1}{8} \bigcirc 2.75$

(c) Compare $6\frac{4}{7}$ and 7.19

$6 \bigcirc 7$ | $6\frac{4}{7} \bigcirc 7.19$

Practice

4 Write >, <, or = in each ◯.

(a) 1.5 ◯ 1.8

(b) 6.52 ◯ 5.62

(c) 1.8 ◯ 1.72

(d) 4.15 ◯ 41.5

(e) 2.3 ◯ $\frac{23}{100}$

(f) $2\frac{1}{4}$ ◯ 2.2

5 Write the numbers in decreasing order.

(a) 8.89, 88.9, 9.88, 8.98

(b) 0.57, $\frac{1}{3}$, 0.77, $1\frac{2}{3}$

6 Write the numbers in increasing order.

(a) 7.4, 7.04, 4.7, 7.7

(b) 132.89, 321.98, 132.98, 132.8

(c) 2.7, $2\frac{4}{5}$, $2\frac{3}{4}$, 2.07

Challenge

7 Use all of the following digits and the decimal point.

6, 1, 5, .

(a) Write the greatest number less than 100.

(b) Write the least number.

Exercise 9

Basics

1

6.26 ↓

|———|———|———|———|———|———|———|———|———|———|
6.2   6.25   6.3

(a) Which number is 6.26 closer to, 6.2 or 6.3?

(b) 6.26 is _____ when rounded to the nearest tenth.

(c) Look at the digit in the _____ place to round to the nearest tenth.

2

6.26 ↓

|———|———|———|———|———|———|———|———|———|———|
6   6.5   7

(a) Which number is 6.26 closer to, 6 or 7?

(b) 6.26 is _____ when rounded to a whole number.

(c) Look at the digit in the _____ place to round to a whole number.

3

12.7 ↓

|———|———|———|———|———|———|———|———|———|———|
12   12.5   13

12.7 is _____ when rounded to a whole number.

12-9 Rounding Decimals   73

4

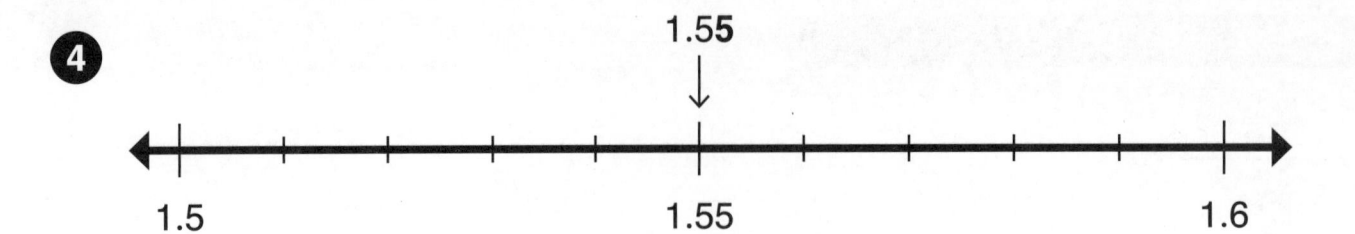

1.55
↓

|———|———|———|———|———|———|———|———|———|———|
1.5   1.55   1.6

1.55 is halfway between 1.5 and _____, so we round up.

1.55 is _____ when rounded to 1 decimal place.

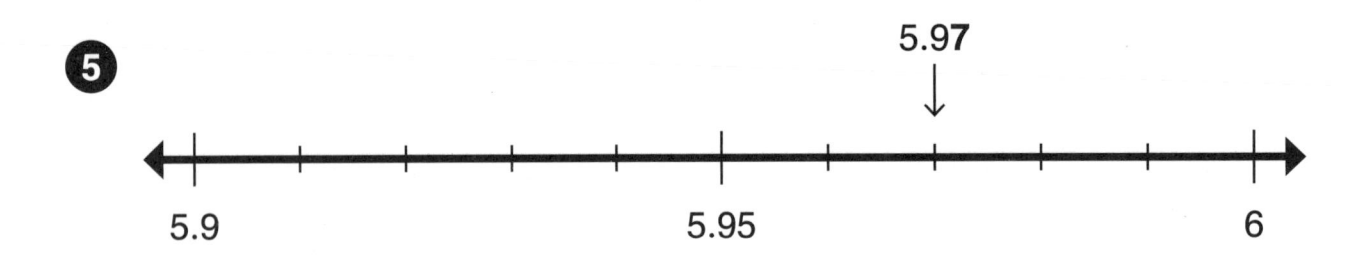

5

5.97
↓

|———|———|———|———|———|———|———|———|———|———|
5.9   5.95   6

5.97 is _____ when rounded to 1 decimal place.

Practice

6 Answer using the given decimals above the number line.

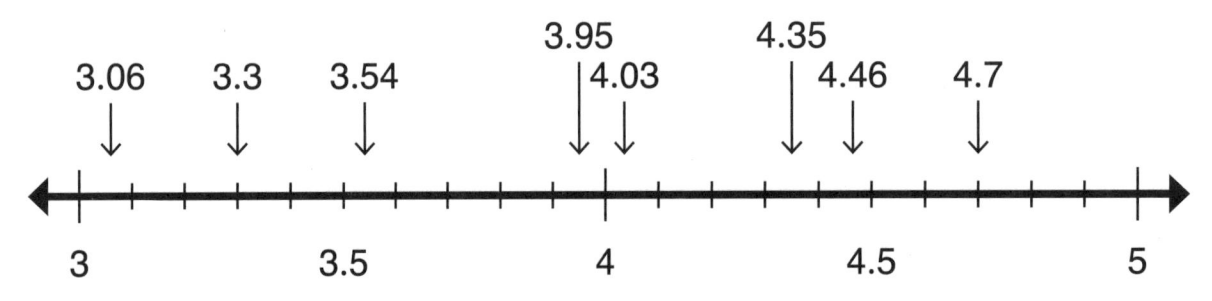

```
                              3.95      4.35
    3.06    3.3    3.54      │ 4.03    │ 4.46   4.7
      ↓      ↓      ↓         ↓  ↓      ↓  ↓     ↓
|—|—|—|—|—|—|—|—|—|—|—|—|—|—|—|—|—|—|—|—|
3              3.5              4             4.5              5
```

(a) Which decimals are 3 when rounded to a whole number?

(b) Which decimals are 4 when rounded to a whole number?

(c) Which decimals are 4.5 when rounded to 1 decimal place?

(d) Which decimals are 4.0 when rounded to 1 decimal place?

7 Round each decimal to 1 decimal place and to a whole number.

Decimal	1 Decimal Place	Whole Number
6.89		
0.64		
9.65		
402.87		
36.04		

Challenge

8 Fill in the missing digit so that the number rounds to 2 when rounded to a whole number and 2.5 when rounded to 1 decimal place.

2 . ☐ 5

9 What is the least number with two decimal places that rounds to 2 when rounded to the nearest tenth?

Exercise 10

Check

1 Complete the number patterns.

(a) | 3.08 | 3.1 | 3.12 | | | |

(b) | 11.07 | 11.05 | 11.03 | | | |

(c) | 19.3 | 19.6 | 19.9 | | | |

2 Write > or < in each ◯.

(a) 16.4 ◯ 14.8

(b) 12.02 ◯ 12.20

(c) 2.3 ◯ $\frac{23}{50}$

(d) $2\frac{7}{40}$ ◯ 2.7

3 Write the numbers in increasing order.

(a) 6.07, 7.06, 6.7, 7.6

(b) 2.92, $2\frac{3}{5}$, $\frac{29}{10}$, $2\frac{43}{50}$

4 A monarch butterfly's wingspan measures 8.93 cm.

  (a) Round this number to a tenth of a centimeter.

  (b) Round this number to a whole number of centimeters.

5 Consider the following numbers.

7.08, 6.98, 7.14, 6.03, 7.03

  (a) Which number is closest to 7?

  (b) Which number is closest to 7.1?

6 Use all of the digits and a decimal point for each number.

3, 4, 5, .

  (a) Write a number that is between 3.35 and 3.53.

  (b) Write two numbers that round to 3.5 when rounded to 1 decimal place.

  (c) Write two numbers that round to 4 when rounded to a whole number.

7 What decimal is the same distance from 6.3 as it is from 6.4 on the number line?

Challenge

8 List all the numbers with 1 decimal place that round to 5 when rounded to a whole number.

9 List all the numbers with two decimal places that round to 3.4 when rounded to 1 decimal place.

Chapter 13 Addition and Subtraction of Decimals

Exercise 1

Basics

1 6 tenths + 3 tenths = ☐ tenths

0.6 + 0.3 = ☐

2 6 tenths + 4 tenths = ☐ tenths = ☐ one

0.6 + 0.4 = ☐

3 6 tenths + 7 tenths = ☐ tenths = ☐ one ☐ tenths

0.6 + 0.7 = ☐

4 8 tenths − 6 tenths = ☐ tenths

0.8 − 0.6 = ☐

5 1 one − 6 tenths = ☐ tenths − 6 tenths = ☐ tenths

1 − 0.6 = ☐

6 1 one 4 tenths − 6 tenths = ☐ tenths − 6 tenths = ☐ tenths

1.4 − 0.6 = ☐

13-1 Adding and Subtracting Tenths 79

Practice

7 Add or subtract.

(a) 0.3 + 0.4 = [ ]

(b) 0.5 + 0.6 = [ ]

(c) 0.9 + 0.8 = [ ]

(d) 0.7 + 0.6 = [ ]

(e) 0.7 − 0.5 = [ ]

(f) 1.7 − 0.9 = [ ]

(g) 1.2 − 0.7 = [ ]

(h) 1.4 − 0.8 = [ ]

8 Write the missing values.

(a) 0.5 + [ ] = 1.2

(b) [ ] + 0.6 = 1.8

(c) 1.3 − [ ] = 0.5

(d) [ ] − 0.6 = 0.4

Challenge

9 Add or subtract using mental math.

(a) 7.8 + 0.9 = [ ]

(b) 4.8 + 0.8 = [ ]

(c) 6.1 − 0.9 = [ ]

(d) 9.4 − 0.7 = [ ]

(e) 0.2 + 0.5 + 0.6 + 0.9 + 0.8 + 0.4 + 0.5 + 0.1 + 0.7 = [ ]

Exercise 2

Basics

1 (a) Estimate the sum of 4.6 and 6.8.

4.6 + 6.8 ≈ 5 + 7 = ☐

(b) Find the sum of 4.6 and 6.8. ⟶

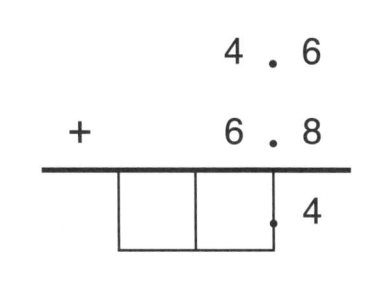

```
     1
     4 . 6
 +   6 . 8
 ┌──┬──┐
 │  │  . 4
 └──┴──┘
```

2 (a) Estimate the sum of 45 and 6.8.

(b) Find the sum of 45 and 6.8. ⟶

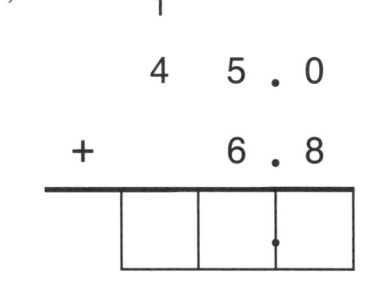

```
     1
   4 5 . 0
 +   6 . 8
 ┌──┬──┐
 │  │  .
 └──┴──┘
```

3 (a) Estimate the sum of 24.7 and 9.5.

(b) Find the sum of 24.7 and 9.5. ⟶

```
 ┌──┬──┐
 │  │  │
 └──┴──┘
   2 4 . 7
 +   9 . 5
 ┌──┬──┐
 │  │  .
 └──┴──┘
```

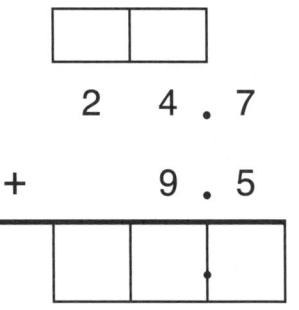

13-2 Adding Tenths with Regrouping

Practice

4 Add. Remember to estimate first.

(a) 9.3 + 6.8

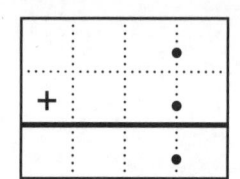

(b) 8.3 + 16.8

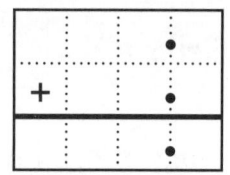

(c) 15.8 + 3.3

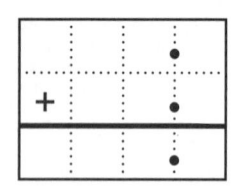

(d) 5.4 + 3.6

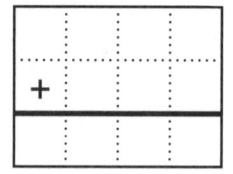

(e) 82 + 9.4

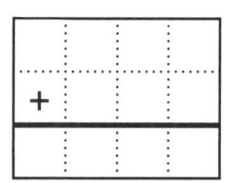

(f) 6.9 + 5.3 + 2.6

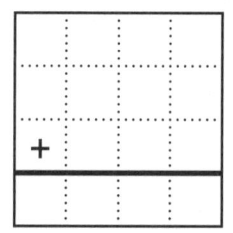

5 To make a dye, Aurora mixed 0.6 L of vinegar with 1.8 L of beet juice and 1.3 L of choke cherry juice. How many liters of red dye did she make?

Exercise 3

Basics

1 (a) Estimate the value of 8.2 − 2.8.

   8.2 − 2.8 ≈ 8 − 3 = ☐

   (b) Subtract 2.8 from 8.2. ⟶

   ```
        7  12
        8 . 2
    −   2 . 8
        ___
          . 4
   ```

2 (a) Estimate the value of 47 − 6.8.

   (b) Subtract 6.8 from 47. ⟶

   ```
        4  7 . 0
    −      6 . 8
   ```

3 (a) Estimate the value of 24.7 − 9.5.

   (b) Subtract 9.5 from 24.7 ⟶

   ```
        2  4 . 7
    −      9 . 5
   ```

13-3 Subtracting Tenths with Regrouping

Practice

4 Subtract. Remember to estimate first.

(a) 9.3 − 6.8

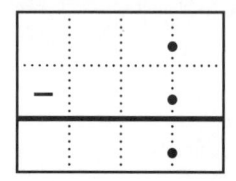

(b) 18.3 − 4.8

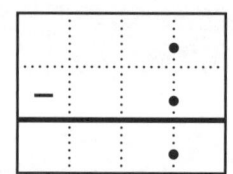

(c) 25.8 − 17.3

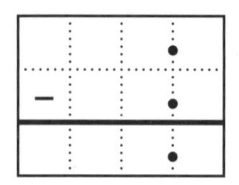

(d) 21.9 − 5.3

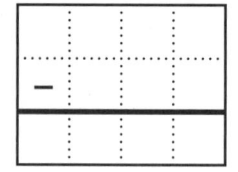

(e) 51.4 − 3

(f) 82 − 9.4

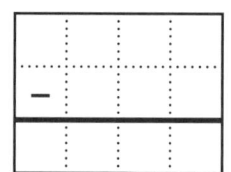

5 A car's gas tank can hold 15 gallons of gas. It had 1.2 gallons of gas. Then Paul added 11.3 gallons. How many more gallons of gas must be added for it to be full?

6 Jada jogged 3.5 km on Saturday. On Sunday she jogged 1.2 km less than on Saturday. On Saturday, Sunday, and Monday she jogged a total of 8.4 km. How far did she jog on Monday?

Challenge

7 There are two strips of paper, one of which is 20.5 cm long. When the two strips of paper are glued together with an overlap of 6.8 cm, the total length is 40 cm. How long is the other strip of paper?

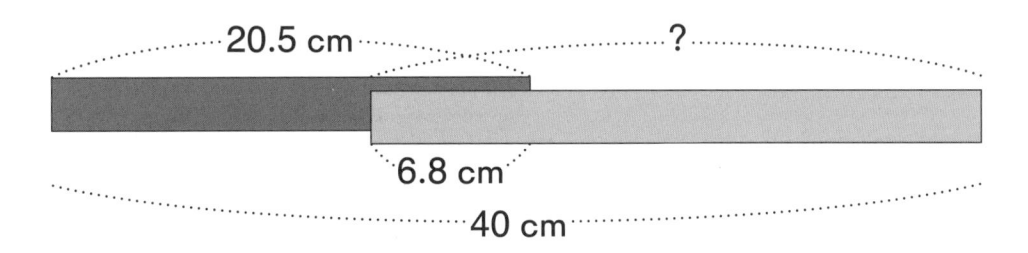

20.5 cm ········ ? ········
······ 6.8 cm ······
······ 40 cm ······

Exercise 4

Check

1 Add or subtract.

(a) 7.8 + 6.8   (b) 9.2 − 6.8   (c) 14.8 + 7.3

(d) 45.2 − 6.7   (e) 74.7 + 5.3   (f) 12.9 − 8

(g) 112.6 + 5.8   (h) 12 − 2.1   (i) 10.2 + 28.7 + 5.5

2 Express the sum of $6\frac{1}{2}$, 4.2, $\frac{3}{5}$, and 8.2 as a decimal.

3 An airline charges an extra baggage fee if a suitcase weighs more than 50 lb. A suitcase by itself weighs 12.8 lb. How much weight can be added to the suitcase so it will still be 5 lb under 50 lb?

4 Jasper weighed himself on the scale and found that he weighed 128.6 lb. Then he weighed both himself and his suitcase and found that together they weighed 185.3 lb. Does the suitcase weigh more or less than 50 lb, and by how much?

5 The capacity of a bottle is 4.2 L. It has $1\frac{3}{5}$ L of water in it now. How much water must be added to fill it? Express the answer as a decimal.

Exercise 5

Basics

1 (a) Estimate the sum of 0.78 and 0.23.

0.78 + 0.23 ≈ 0.8 + 0.2 = ☐

(b) Find the sum of 0.78 and 0.23. ⟶

```
      1   1
      0 . 7   8
  +   0 . 2   3
  ─────────────
      ☐ . ☐   1
```

2 (a) Estimate the sum of 5.65 and 6.8.

5.65 + 6.8 ≈ 6 + 7 = ☐

(b) Find the sum of 5.65 and 6.8. ⟶

```
          ☐
      5 . 6   5
  +   6 . 8   0
  ─────────────
    ☐ ☐ . ☐   ☐
```

3 (a) Estimate the sum of 0.81 and 29.5.

0.81 + 29.5 ≈ 1 + 30 = ☐

(b) Find the sum of 0.81 and 29.5. ⟶

```
        ☐   ☐
    2   9 . 5   0
  +     0 . 8   1
  ───────────────
    ☐   ☐ . ☐   ☐
```

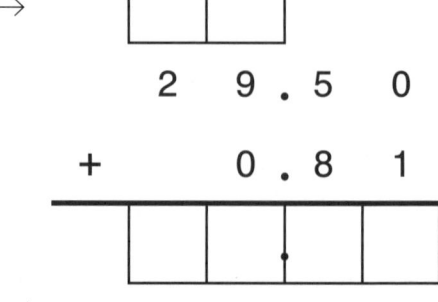

13-5 Adding Hundredths

Practice

4 Add. Remember to estimate first.

(a) 0.65 + 0.85

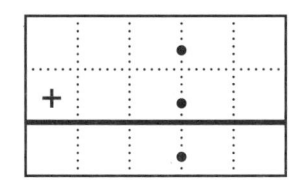

(b) 8.32 + 16.8

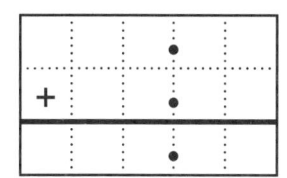

(c) 15.81 + 7.32

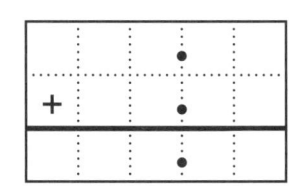

(d) 8.95 + 0.67

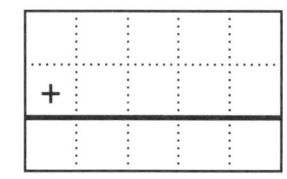

(e) 73 + 9.48

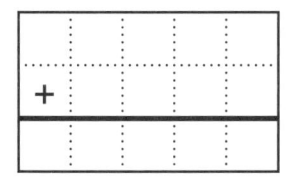

(f) 2.56 + 5.3 + 0.62

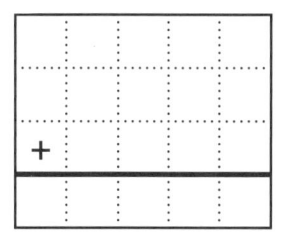

5 In a laboratory, 3 chemicals were weighed and then mixed together before adding water to the mixture. Chemical A weighed 6.91 g, Chemical B weighed 7.37 g, and Chemical C weighed 0.80 g. What is the total weight of the three chemicals?

Exercise 6

Basics

1 Cross off discs to show the answer.

(a) Subtract 7 from 10.

10 − 7 = ☐

(b) Subtract 0.07 from 0.1.

1 tenth − 7 hundredths = ☐ hundredths − 7 hundredths

= ☐ hundredths

0.1 − 0.07 = ☐

(c) Subtract 0.07 from 0.2.

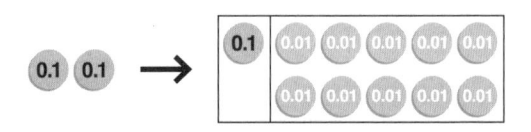

0.2 − 0.07 = ☐

2 (a) 50 + ☐ = 90   (b) 0.5 + ☐ = 0.9

(c) 3 + ☐ = 10   (d) 0.03 + ☐ = 0.1

(e) 53 + ☐ = 100   (f) 0.53 + ☐ = 1

(g) 100 − 53 = ☐   (h) 1 − 0.53 = ☐

(i) 200 − 53 = ☐   (j) 2 − 0.53 = ☐

13-6 Subtracting from 1 and 0.1

Practice

3 (a) 0.1 − 0.09 = ☐   (b) 0.1 − 0.04 = ☐

(c) 0.1 − 0.02 = ☐   (d) ☐ + 0.05 = 0.1

(e) 0.5 − 0.06 = ☐   (f) 0.4 − 0.01 = ☐

4 (a) 1 − 0.85 = ☐   (b) 1 − 0.89 = ☐

(c) 1 − 0.44 = ☐   (d) 1 − 0.72 = ☐

(e) 0.17 + ☐ = 1   (f) ☐ + 0.28 = 1

(g) 3 − 0.25 = ☐   (h) 4 − 0.33 = ☐

Challenge

5 Add or subtract using mental math.

(a) 3.76 + 0.99 = ☐

(b) 7.87 + 0.98 = ☐

(c) 3.7 + 0.98 = ☐

(d) 1.5 − 0.98 = ☐

(e) 5.22 + 3.99 = ☐

(f) 7.07 − 3.98 = ☐

13-6 Subtracting from 1 and 0.1

Exercise 7

Basics

1 (a) Estimate the value of 4.13 − 0.68.

   4.13 − 0.68 ≈ 4 − 0.7 = ☐

   (b) Subtract 0.68 from 4.13. ⟶

```
        3  10  13
        4 . 1   3
    −   0 . 6   8
    ─────────────
        ☐ . ☐   5
```

2 (a) Estimate the value of 23.46 − 7.8.

   (b) Subtract 7.8 from 23.46. ⟶

```
        ☐   ☐   ☐
        2   3 . 4   6
    −       7 . 8   0
    ───────────────────
        ☐   ☐ . ☐   ☐
```

3 (a) Estimate the value of 42.7 − 2.62.

   (b) Subtract 2.62 from 42.7. ⟶

```
            ☐   ☐
        4   2 . 7   0
    −       2 . 6   2
    ───────────────────
        ☐   ☐ . ☐   ☐
```

Practice

4 Subtract. Remember to estimate first.

(a) 0.61 − 0.38

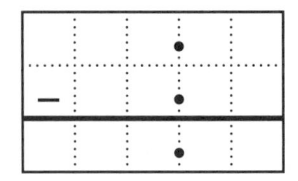

(b) 8.32 − 0.49

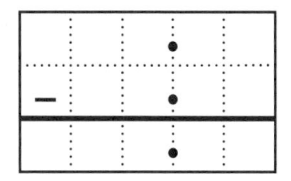

(c) 15.18 − 7.32

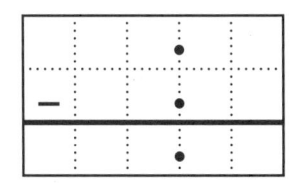

(d) 8.35 − 0.6

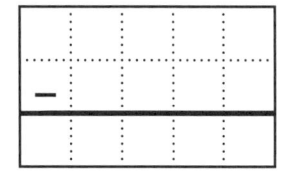

(e) 73 − 9.48

(f) 5.03 − 4.56

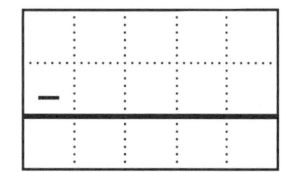

5 A beaker is filled with 1.2 L of a salt solution. 0.37 L of the solution was poured out from it. How many liters of solution are still in the beaker?

6. In a science experiment, students hung a 100 g weight on a string and measured the time it took for a pendulum to swing back and forth once.

Trial	Length of string (m)	Time (s)
A	0.15	0.83
B	0.5	1.5
C	0.9	1.93

(a) What was the difference in the length of string between Trial A and Trial B and between Trial B and Trial C?

(b) What is the difference in time between Trial A and Trial B and between Trial B and Trial C?

(c) Did the time increase or decrease as the length of the string increased?

(d) What is the length of string in centimeters for each trial?

Exercise 8

Basics

1 Complete the table to express the value of each coin in dollars and as a fraction of $1 in simplest form.

Coin	Dollars	Fraction of $1
Penny	$0.01	$\frac{1}{100}$
Nickel		
	$0.10	
		$\frac{25}{100} = \boxed{\frac{}{}}$
50-cent coin		

2 Add $4.75 and $6.35.

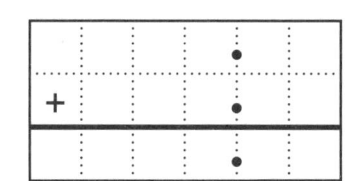

3 Clara had $10. She then spent $4.27. How much money does she have left?

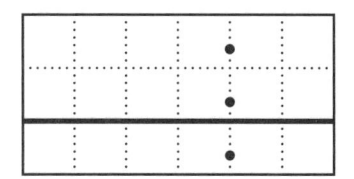

13-8 Money, Decimals, and Fractions 95

Practice

4 Express each of the following in dollars and as a fraction of $1 in simplest form.

	Dollars	Fraction of $1
15¢		
52¢		
7 nickels		
2 quarters, 5 nickels		
1 quarter, 1 dime, 1 penny		

5 Add or subtract.

(a) $11.56 − $5.56

(b) $81.35 − $9

(c) $6.21 + $0.38

(d) $87.32 + $0.69

(e) $20.18 + $7.32

(f) $42 − $9.48

6 Franco had $2. He spent $1.60. What fraction of the money does he have left?

7 Aisha bought two books and a bookmark. One book cost $5.45, the other book cost $8.28, and the bookmark cost $1.99. She paid with a $20 bill. How much change did she receive?

Challenge

8 Hannah has $34.50 more than Clemens. Clemens has $132.30. Paula has $14.80 less than Clemens. They want to combine their money to buy a raft that costs $449.29. How much more money do they need?

Exercise 9

Check

1 (a) $0.1 - 0.07 = \boxed{}$ (b) $1 - 0.06 = \boxed{}$

(c) $1 - 0.28 = \boxed{}$ (d) $1 - 0.44 = \boxed{}$

(e) $0.08 + \boxed{} = 0.1$ (f) $\boxed{} + 0.19 = 1$

(g) $5 - 0.85 = \boxed{}$ (h) $10 - 0.33 = \boxed{}$

2 Express each of the following in dollars and as a fraction of $1 in simplest form.

	Dollars	Fraction of $1
30¢		
92¢		
3 dimes, 3 nickels		
1 quarter, 5 nickels, 6 pennies		
2 quarters, 1 dime, 5 pennies		

3 75¢ is what fraction of $5?

4 Follow the arrows and fill in the missing numbers. Write + or − in each ◯.

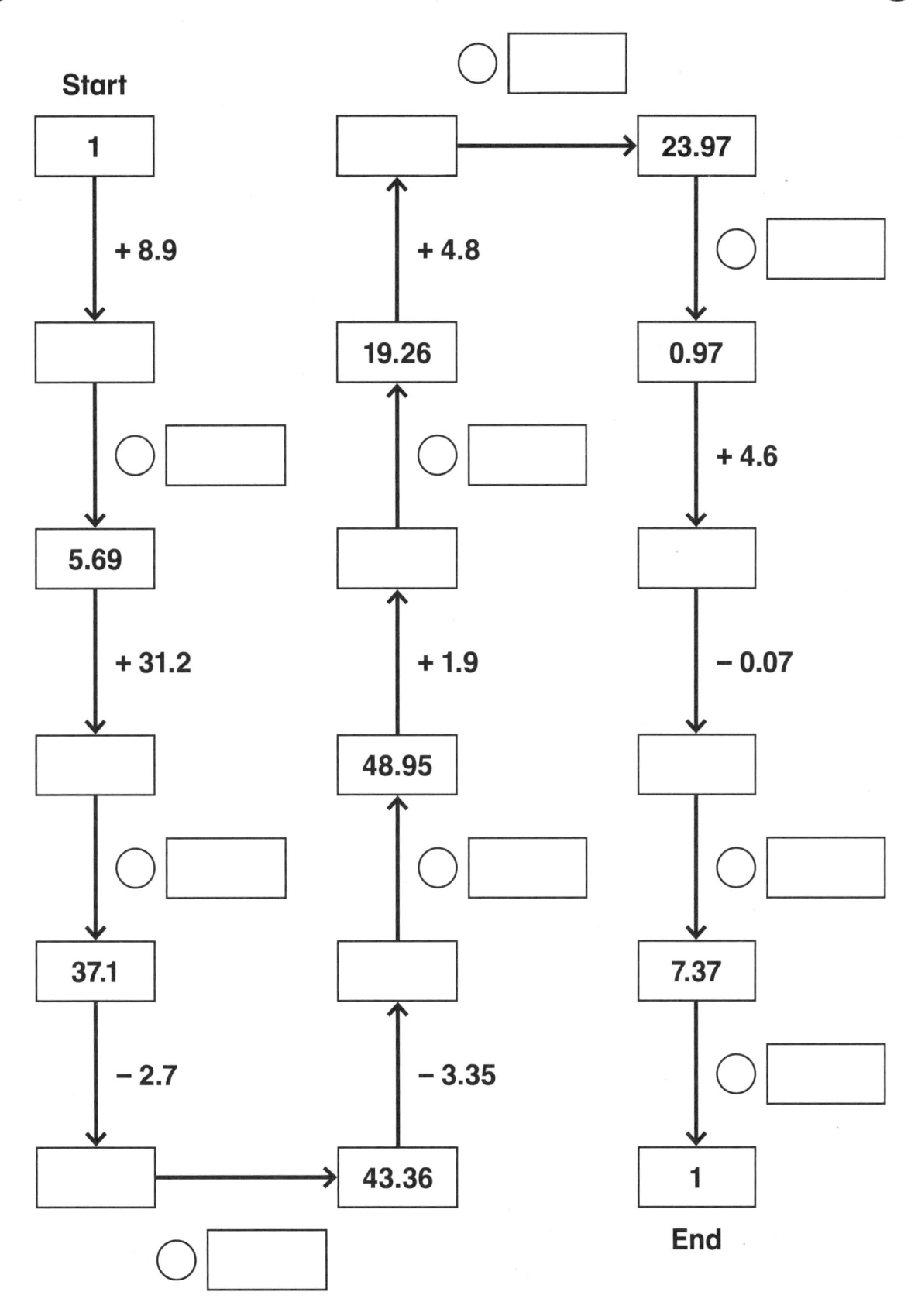

Start: 1

1 → +8.9 → [9.9] ◯[−] [4.21] → 5.69 → +31.2 → [36.89] ◯[+] [0.21] → 37.1 → −2.7 → [34.4] → 43.36

43.36 ◯[+] [__] (from below start path continuation)

43.36 → −3.35 → [40.01] ...

48.95 → +1.9 → [50.85] ◯[−] [__] → 48.95

19.26 ← +4.8 ← [14.46] ...

Top: [28.77] ◯[−] [__] → 23.97

23.97 ◯[−] [23] → 0.97 → +4.6 → [5.57] → −0.07 → [5.5] ◯[−] [__] → 7.37 ◯[−] [6.37] → 1

End: 1

5 Express the sum of $6\frac{7}{20}$, 2.4, $\frac{8}{25}$, and 4.12 as a decimal.

6 Bron bought a pad of watercolor paper that cost $17.30 and a set of watercolor paints. The paints cost $4.31 more than the paper. He paid with two $20 bills. How much change did he receive?

7 Tiara grew three pumpkins in different areas of her yard. She weighed the pumpkins 3 weeks before she harvested them, and then again the day she harvested them. The total weight of the three pumpkins when she harvested them was 28.7 kg. The table shows the results. Complete the table. Which pumpkin grew the most over the 3 weeks?

	Starting weight	Weight after 3 weeks	Difference in weight
Pumpkin A	8.6 kg	9.24 kg	
Pumpkin B	7.26 kg	10.89 kg	
Pumpkin C	6.74 kg		

Exercise 10

Check

1 | 854.6 | 89.05 | 70.54 | 504.2 |

(a) What does the digit 5 stand for in each number?

(b) Write each number in expanded form.

(c) Write each number as a mixed number in simplest form.

2 Write >, <, or = in each ◯.

(a) $4\frac{3}{5} - 1\frac{1}{10}$ ◯ $10.6 - 7.4$

(b) $12 - 10.2$ ◯ $8 \times \frac{7}{20}$

(c) $13.35 - 9.6$ ◯ $1\frac{1}{2} + 2\frac{1}{4}$

(d) $\frac{5}{6} \times 18$ ◯ $10.55 + 5.68$

(e) 7 m 8 cm + 4 m 20 cm ◯ 7.8 m + 4.2 m

Review 3 101

3. This table shows the rainfall, in inches, for each month over one year in a city.

Month	Jan	Feb	Mar	Apr	May	Jun	Jul	Aug	Sep	Oct	Nov	Dec
Rainfall (in)	4.7	3	3.2	2.7	2.5	1.5	1.2	1.2	1.8	3.7	5.8	4.2

(a) Graph this information below.

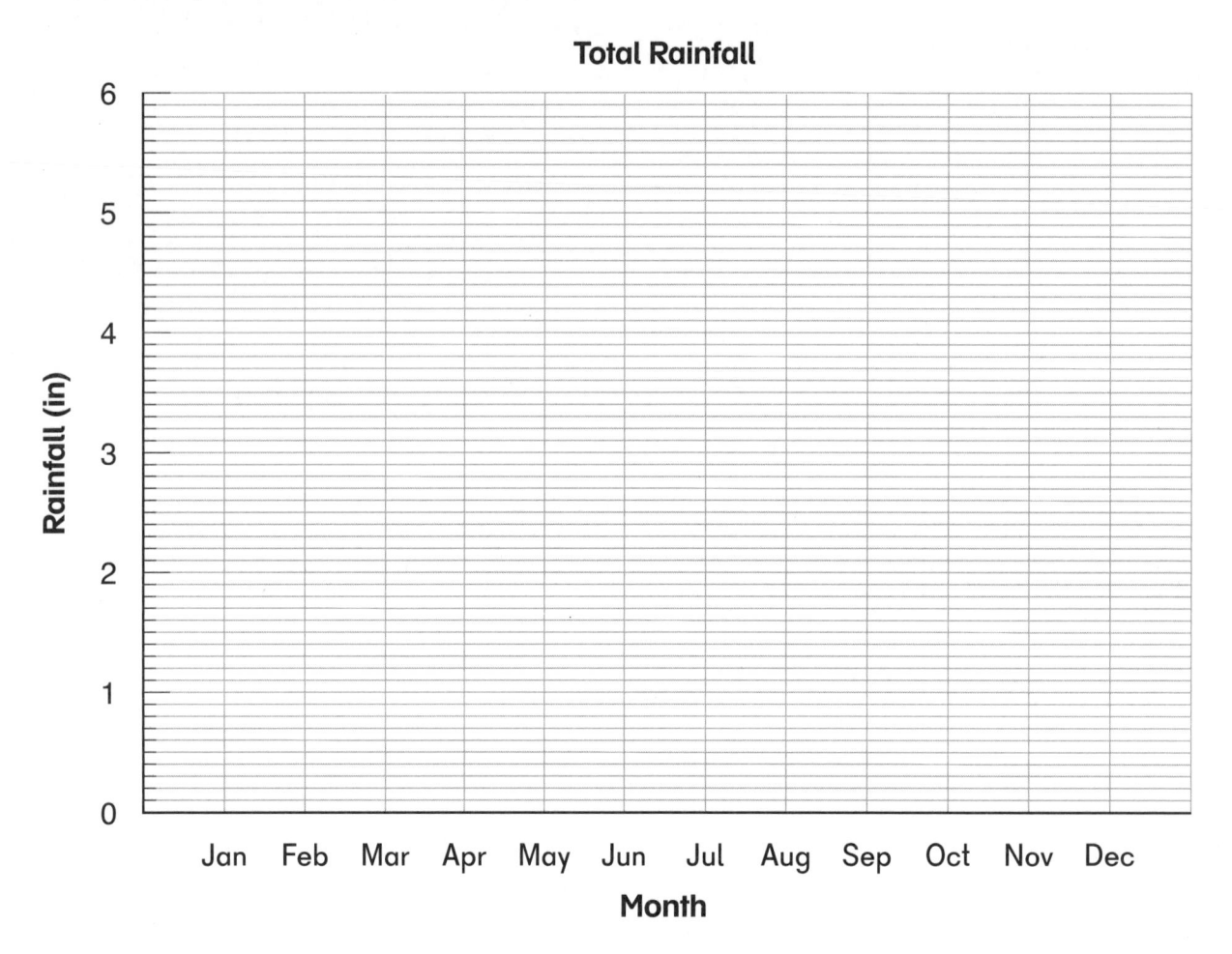

Total Rainfall

(b) Which month had the most rainfall?

(c) Between which months was there a monthly increase in rainfall?

(d) Which months had the same amount of rainfall?

(e) What was the total amount of rainfall in inches for the year?

(f) The total amount of rainfall for the year is how many inches less than 3 feet?

(g) What fraction of the months had more than $1\frac{1}{4}$ inches of rain?

4. The perimeter of a rectangular picture frame is 100 in. The width of the frame is 1 ft 8 in.

(a) Express the length of the frame in feet and inches.

(b) Express the length of the frame in feet as a mixed number.

(c) Express the length of the frame in feet as a decimal.

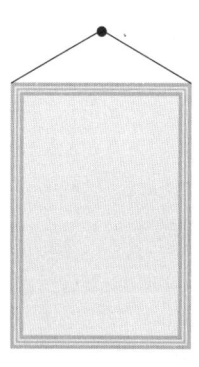

5. Find the area in square centimeters and perimeter in centimeters of the following figure.

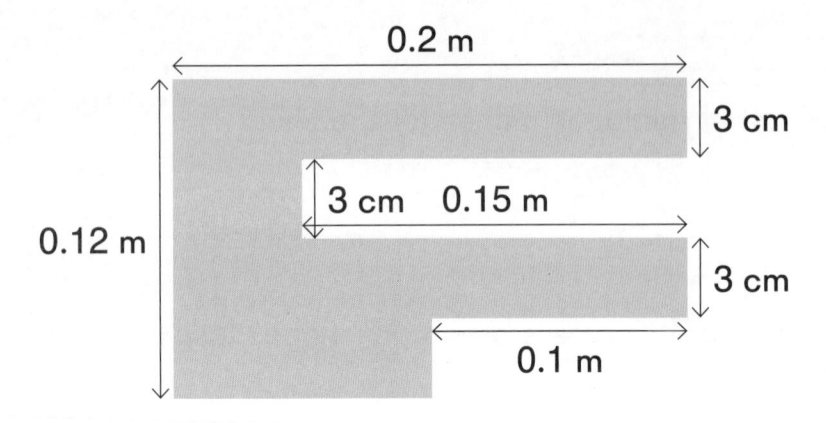

```
        0.2 m
    ←──────────→
    ┌──────────┐ ↕ 3 cm
    │          │
    │  3 cm  0.15 m
0.12 m  ←──→←────────→
    │   ┌──────────┐ ↕ 3 cm
    │   │          │
    │   └──────────┘
    │       0.1 m
    └───→←────────→
```

6. Aaron needs 1,000 yd of paracord, a type of rope, to make a hammock. He has two 1,000-ft spools of paracord, three 250-ft spools of paracord, and a 20-yd length of paracord. How many feet of paracord does he still need?

7 There are between 50 and 90 children at a sports event. They can be divided into 6 equal groups or 8 equal groups. How many children are there?

8 A bin of nails weighs 14 kg 210 g. The bin by itself weighs 2 kg 300 g. 4 kg 30 g of nails were removed. What is the weight of the nails left in the bin? Express the answer in compound units.

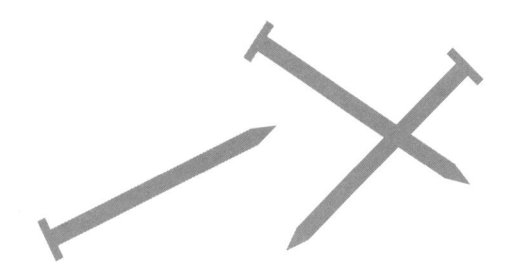

9 Violet spent $\frac{1}{3}$ of her money on a book and $3.60 on a snack. She had $12.40 left. How much money did she have to start with?

Challenge

10 Two bags of beans together weighed 15.82 kg. Bag A weighed 7.45 kg. After some beans were transferred from Bag B to Bag A, Bag A weighed 10 kg. What is the weight of Bag B now? Express the answer as a decimal.

11 This square is formed by 16 identical small rectangles. The perimeter of each small rectangle is 10 cm. What is the area of the square?

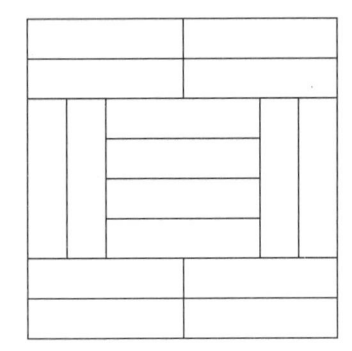

12 Owen has a 3-quart and a 5-quart jar. He wants to measure 1 gal of water. How can he use the jars he has to do this?

Chapter 14 Multiplication and Division of Decimals

Exercise 1

Basics

1 (a) 7 ones × 4 = ☐ ones

7 × 4 = ☐

(b) 7 tenths × 4 = ☐ tenths

0.7 × 4 = ☐

(c) 7 hundredths × 4 = ☐ hundredths

0.07 × 4 = ☐

2 (a) 8 ones × 5 = ☐ ones = ☐ tens

8 × 5 = ☐

(b) 8 tenths × 5 = ☐ tenths = ☐ ones

0.8 × 5 = ☐

(c) 8 hundredths × 5 = ☐ hundredths = ☐ tenths

0.08 × 5 = ☐

Practice

③ (a) 90 × 8 = ☐   (b) 50 × 6 = ☐

   9 × 8 = ☐   5 × 6 = ☐

   0.9 × 8 = ☐   0.5 × 6 = ☐

   0.09 × 8 = ☐   0.05 × 6 = ☐

④ Complete the number patterns.

(a) | 0.7 | 1.4 | 2.1 | ☐ | ☐ | ☐ | ☐ |

(b) | 0.3 | 0.36 | 0.42 | ☐ | ☐ | ☐ | ☐ |

(c) | 0.45 | 0.54 | 0.63 | ☐ | ☐ | ☐ | ☐ |

⑤ (a) 0.3 × 4 = ☐   (b) 0.4 × 3 = ☐

   (c) 0.07 × 8 = ☐   (d) 0.05 × 4 = ☐

   (e) 0.06 × 7 = ☐   (f) 0.04 × 2 = ☐

⑥ A square is 0.8 m on one side. What is its perimeter?

14-1 Multiplying Tenths and Hundredths

Exercise 2

Basics

1 (a) Estimate the product of 6.7 and 8.

6.7 × 8 ≈ 7 × 8 = ☐

(b) Multiply 6.7 by 8.

First multiply the tenths.

0.7 × 8 = ☐

```
    5
  6.7
×   8
─────
   .6
```

Then multiply the ones and add in any renamed tenths.

(6 × 8) + 5 = ☐

6.7 × 8 = ☐☐.6

```
    5
  6.7
×   8
─────
  ☐☐.6
```

2 (a) Estimate the product of 56.5 and 6.

56.5 × 6 ≈ 60 × 6 = ☐

(b) Multiply 56.5 by 6.

```
   ☐ 3
  56.5
×    6
──────
  ☐☐☐.0
```

56.5 × 6 = ☐☐☐.0 = ☐☐☐

14-2 Multiplying Decimals by a Whole Number — Part 1

Practice

3 (a) Alex estimated the product of 26.9 and 4 to be 120. With what number did he replace 26.9?

(b) Sofia estimated the product of 26.9 and 4 to be 100. With what number did she replace 26.9?

(c) Whose estimate will be closer to the actual product? Why?

(d) Find the product of 26.9 and 4.

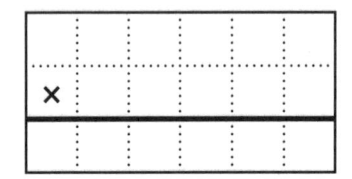

4 Estimate and then find the actual product.

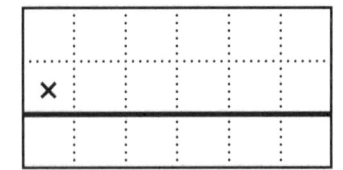

(a) 8.3 × 4 ≈ ☐

(b) 348.2 × 5 ≈ ☐

8.3 × 4 = ☐

348.2 × 5 = ☐

5 Multiply. Remember to estimate first.

(a) 7.8 × 6

(b) 9.2 × 5

(c) 14.8 × 9

(d) 702.5 × 8

6 A brick weighs 3.5 kg. A concrete block weighs 5 times as much as the brick. How much does the concrete block weigh?

7 There are 7 fence posts in a row. The distance between the middle of each post is 34.5 cm. How far is it from the middle of the first post to the middle of the last post? Express the answer in meters and centimeters.

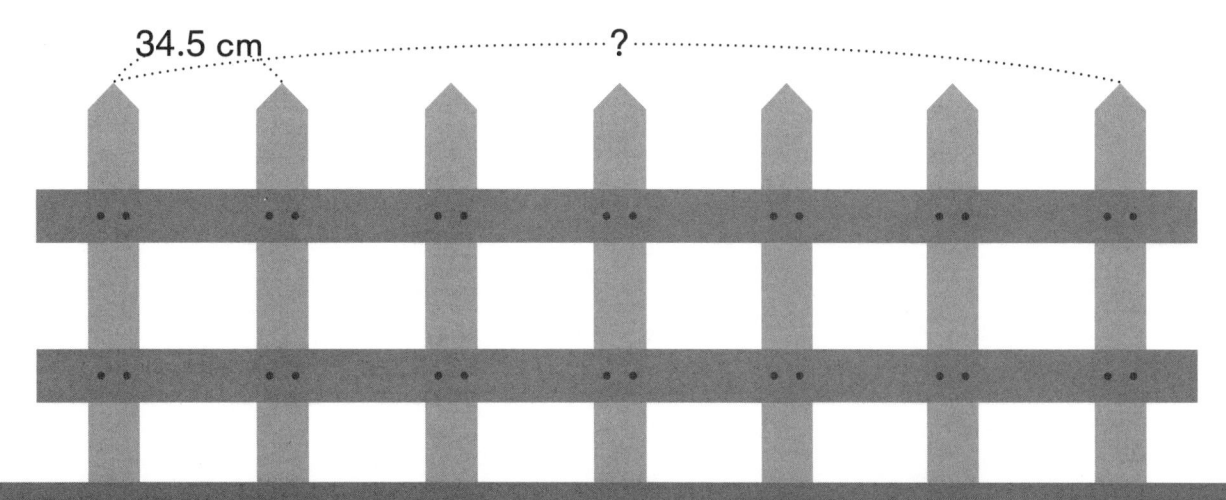

34.5 cm   ?

Exercise 3

Basics

1 (a) Estimate the product of 0.78 and 6.

0.78 × 6 ≈ 0.8 × 6 = ☐

(b) Multiply 0.78 by 6.

First multiply the hundredths.

0.08 × 6 = ☐

```
      4
   0 . 7  8
 ×        6
          8
```

Then, multiply the tenths and add any renamed hundredths.

(0.7 × 6) + 0.4 = ☐

0.78 × 6 = ☐.☐ 8

```
      4
   0 . 7  8
 ×        6
   ☐ .☐  8
```

2 (a) Estimate the product of 4.78 and 5.

4.78 × 5 ≈ 5 × 5 = ☐

(b) Multiply 4.78 by 5.

```
    ☐ 4
    4 . 7  8
 ×         5
    ☐ ☐.☐ 0
```

4.78 × 5 = ☐☐.☐ 0 = ☐☐.☐

Practice

3 0.72 × 5 is closest to which of the following?

| 5 | 3.5 | 50 | 35 |

4 9.72 × 3 is closest to which of the following?

| 2.7 | 27 | 29 | 32 |

5 Write > or < in each ◯. Use estimation.

(a) 0.89 × 7 ◯ 1.5 × 3

(b) 4.1 × 8 ◯ 6.29 × 4

6 Estimate and then find the actual product.

(a) 0.56 × 7 ≈ ☐   (b) 45.72 × 5 ≈ ☐

0.56 × 7 = ☐   45.72 × 5 = ☐

14-3 Multiplying Decimals by a Whole Number — Part 2

7 Multiply. Remember to estimate first.

(a) 7.18 × 6

(b) 0.82 × 3

(c) 4.18 × 9

(d) 74.71 × 8

8 Renata has $20. She wants to buy 8 gel pens that cost $2.99 each. Does she have enough money?

9 One t-shirt costs $8.99. A package of 5 t-shirts costs $43. Is it more or less expensive to buy 5 individual t-shirts and by how much?

Exercise 4

Check

1 (a) 0.6 × 4 = ☐   (b) 0.08 × 3 = ☐

   (c) 0.07 × 9 = ☐   (d) 0.05 × 2 = ☐

   (e) 0.05 × 5 = ☐   (f) 0.8 × 4 = ☐

2 Estimate to arrange the expressions in order from least to greatest.

0.68 × 9	1.4 × 3	12.3 × 2	0.34 × 8
A	B	C	D

3 Multiply.

   (a) 2.82 × 6   (b) 12.2 × 5

   (c) 0.95 × 5   (d) 7.12 × 3

   (e) 23.27 × 7   (f) 444.4 × 8

14-4 Practice A

4 A 2-inch ball bearing costs $9.45 and a 1-inch ball bearing costs $0.77. Mr. Jung bought eight 1-inch ball bearings and two 2-inch ball bearings. How much did he spend?

5 A rectangular field is 0.26 km long and 0.08 km wide. What is the perimeter of the field?

Challenge

6 Julian has $14.85. Hudson has 4 times as much money as Julian. Simone has $5.60 less than twice as much money as Julian. How much money do they have altogether?

Exercise 5

Basics

1 (a) 24 ones ÷ 8 = [3] ones

24 ÷ 8 = [3]

(b) 24 tenths ÷ 8 = [3] tenths

2.4 ÷ 8 = [0.3]

(c) 24 hundredths ÷ 8 = [3] hundredths

0.24 ÷ 8 = [0.03]

2 (a) 56 ÷ 8 = [7] ones ÷ 8 = [ ] ones = [ ]

(b) 5.6 ÷ 8 = [0.7] tenths ÷ 8 = [ ] tenths = [ ]

(c) 0.56 ÷ 8 = [0.07] hundredths ÷ 8 = [ ] hundredths = [ ]

3 (a) 720 ÷ 8 = [90]

72 ÷ 8 = [9]

7.2 ÷ 8 = [0.9]

0.72 ÷ 8 = [0.09]

(b) 300 ÷ 5 = [60]

30 ÷ 5 = [6]

3 ÷ 5 = [0.6]

0.3 ÷ 5 = [0.06]

14-5 Dividing Tenths and Hundredths

Practice

4 (a) $0.4 \div 2 =$ ☐   (b) $0.2 \div 4 =$ ☐

(c) $0.09 \div 3 =$ ☐   (d) $3.6 \div 9 =$ ☐

(e) $0.36 \div 6 =$ ☐   (f) $4.2 \div 7 =$ ☐

(g) $8.1 \div 9 =$ ☐   (h) $0.3 \div 6 =$ ☐

5 The perimeter of a square tile is 3.6 ft. How long is one side? Express the answer as a decimal.

6 A package of 6 plates costs $3. What is the cost of 1 plate?

Exercise 6

Basics

1 (a) Complete the following estimations for 93.6 ÷ 4.

80 ÷ 4 = ☐

100 ÷ 4 = ☐

The quotient will be between _____ and _____.

(b) Divide 93.6 by 4.

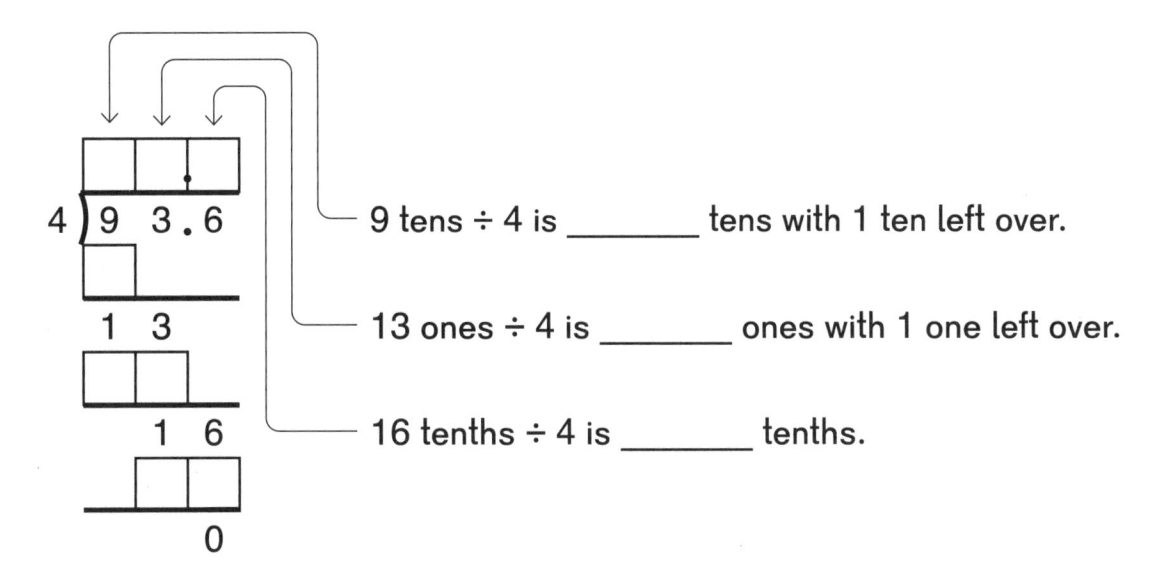

4) 9 3 . 6

9 tens ÷ 4 is _____ tens with 1 ten left over.

13 ones ÷ 4 is _____ ones with 1 one left over.

16 tenths ÷ 4 is _____ tenths.

(c) Compare the estimates to the actual product.

Which estimate is lower?

Which estimate is higher?

Which estimate is closest?

(d) Check: ☐ × 4 = 93.6

Practice

2 (a) Sofia estimated 52.2 ÷ 6 by dividing 60 by 6.
Mei estimated 52.2 ÷ 6 by dividing 54 by 6. Find their estimates.

(b) Which estimate will be closer to the actual answer? Why?

(c) Divide 52.2 by 6.

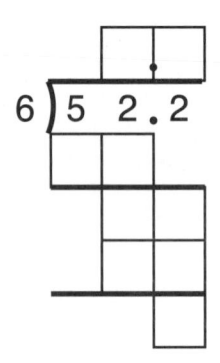

3 Estimate and then find the quotient.

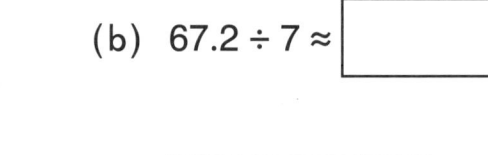

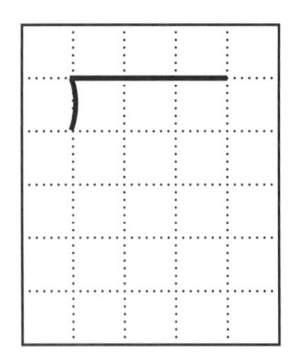

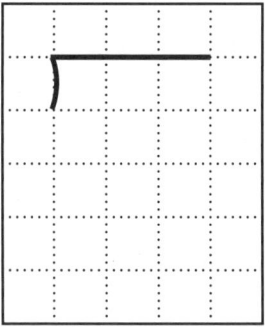

(a) 7.8 ÷ 3 ≈ ☐

(b) 67.2 ÷ 7 ≈ ☐

7.8 ÷ 3 = ☐

67.2 ÷ 7 = ☐

14-6 Dividing Decimals by a Whole Number — Part 1

4 Divide. Remember to estimate mentally first.

(a) 7.6 ÷ 4

(b) 9.5 ÷ 5

(c) 71.6 ÷ 2

(d) 702.4 ÷ 8

5 The capacity of a pail is 7 times as much as the capacity of a bottle. If the capacity of the pail is 9.1 L, what is the capacity of the bottle?

6 6 lampposts are equally spaced in a row. The total distance between the first post and the last post is 75.5 m. What is the distance between each lamppost?

Exercise 7

Basics

1 (a) Complete the following estimations for 8.28 ÷ 6.

6 ÷ 6 = [ ]   12 ÷ 6 = [ ]

The quotient will be between _____ and _____.

(b) Divide 8.28 by 6.

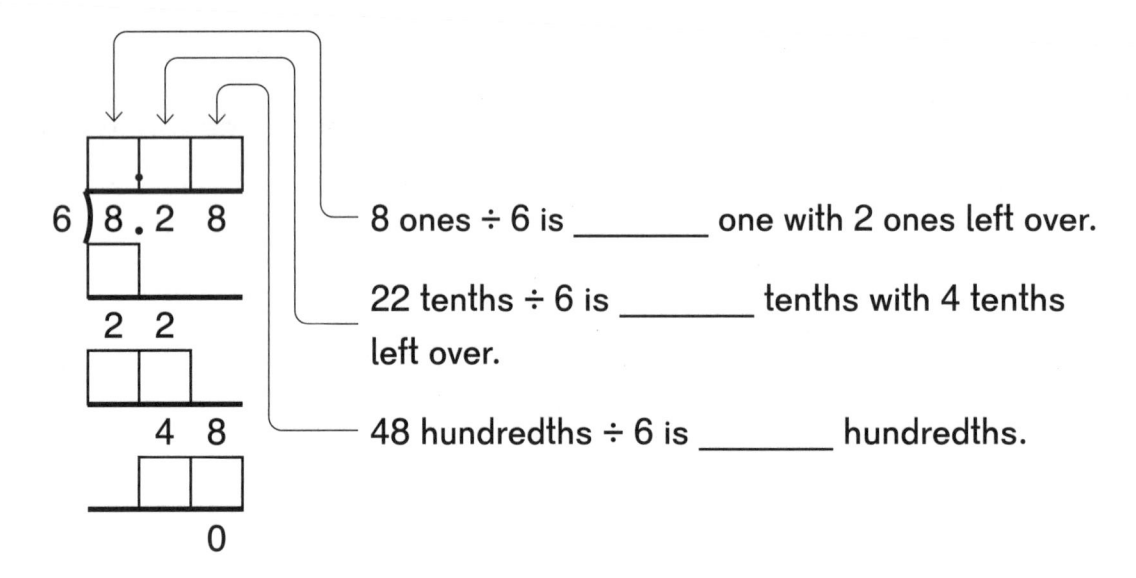

6) 8.28

8 ones ÷ 6 is _____ one with 2 ones left over.

22 tenths ÷ 6 is _____ tenths with 4 tenths left over.

48 hundredths ÷ 6 is _____ hundredths.

(c) Check: [ ] × 6 = 8.28

2 Which of the following have a quotient greater than 1 but less than 10? Use estimation.

| 0.72 ÷ 9 | 4.62 ÷ 2 | 3.76 ÷ 8 | 51.17 ÷ 7 | 52.64 ÷ 4 |

14-7 Dividing Decimals by a Whole Number — Part 2

Practice

3 1.76 ÷ 6 is closest to which number below?

| 0.1 | 0.2 | 0.3 | 0.4 |

4 6.64 ÷ 8 is greater than which numbers below?

| 0.7 | 0.8 | 0.9 | 1 |

5 Estimate and then find the exact quotient.

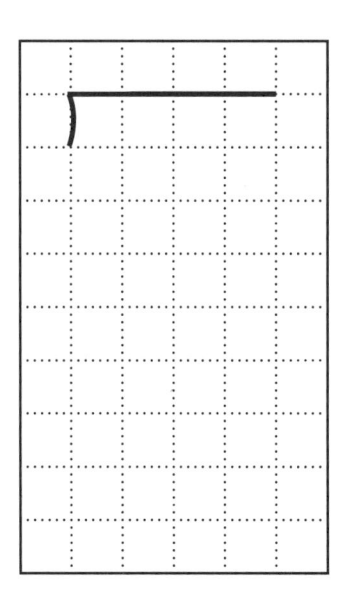

(a) 51.17 ÷ 7 ≈ ☐ (b) 69.45 ÷ 5 ≈ ☐

51.17 ÷ 7 = ☐ 69.45 ÷ 5 = ☐

14-7 Dividing Decimals by a Whole Number — Part 2

6 Divide. Remember to estimate mentally first.

(a) 0.85 ÷ 5

(b) 9.54 ÷ 6

(c) 28.98 ÷ 3

(d) 702.44 ÷ 4

7 A package of five 1-inch steel ball bearings costs $7.95. What is the cost of one ball bearing?

8 A bottle of perfume costs 4 times as much as 3 tubes of toothpaste. The bottle of perfume costs $35.16. What is the cost of one tube of toothpaste?

Exercise 8

Basics

1 (a) Estimate 26 ÷ 8.

26 ÷ 8 ≈ 24 ÷ 8 = ☐

(b) Divide 26 by 8.

8) 2 6 . 0 0

2 (a) Estimate 9.6 ÷ 5.

9.6 ÷ 5 ≈ ☐

(b) Divide 9.6 by 5.

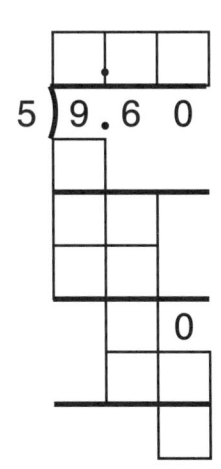

5) 9 . 6 0

Practice

3 Divide. Remember to estimate first.

(a) 45 ÷ 6

(b) 81 ÷ 4

(c) 80.8 ÷ 5

(d) 56.4 ÷ 6

(e) 452 ÷ 5

(f) 91.1 ÷ 2

4 A package of 4 pens cost $19. What is the cost of 1 pen?

Exercise 9

Check

1 (a) 2.4 ÷ 3 = ☐ (b) 0.08 ÷ 4 = ☐

(c) 1 ÷ 5 = ☐ (d) 4.9 ÷ 7 = ☐

(e) 0.4 ÷ 8 = ☐ (f) 0.54 ÷ 9 = ☐

2 Divide.

(a) 8.7 ÷ 3 (b) 9.48 ÷ 6

(c) 2.94 ÷ 7 (d) 13 ÷ 4

(e) 62.2 ÷ 5 (f) 8.1 ÷ 6

3 Three children shared the cost of a birthday present equally. The present cost $29.40. How much did each child contribute?

4 Lisa bought 3 boxes of tissues and 1 package of paper towels for $16.93. The paper towels cost $10.99. What is the price of one box of tissues?

5 Melvin had 12 bags of topsoil. Each bag of topsoil weighed 30.8 pounds. He divided the topsoil equally among 8 plots in his garden. How many pounds of topsoil did he put in each plot?

6 Rope A is 4 times as long as Rope B. Rope B is 14.3 m shorter than Rope C. The total length of the three ropes is 71.9 m. How long is Rope C?

7 4 bowls and 4 plates cost $94.32 altogether. Each plate costs $4.50 more than each bowl. What is the cost of the 4 bowls?

Challenge

8 2 plates and 2 bowls cost $22.20. 4 plates and 1 bowl cost $30. How much does 1 bowl cost?

9 Katie has $40.85 more than Eliza. After Katie spent $12.25, she had 3 times as much money as Eliza. How much money did Katie have at first?

Chapter 15 Angles

Exercise 1

Basics

1 A complete turn is 360 degrees.

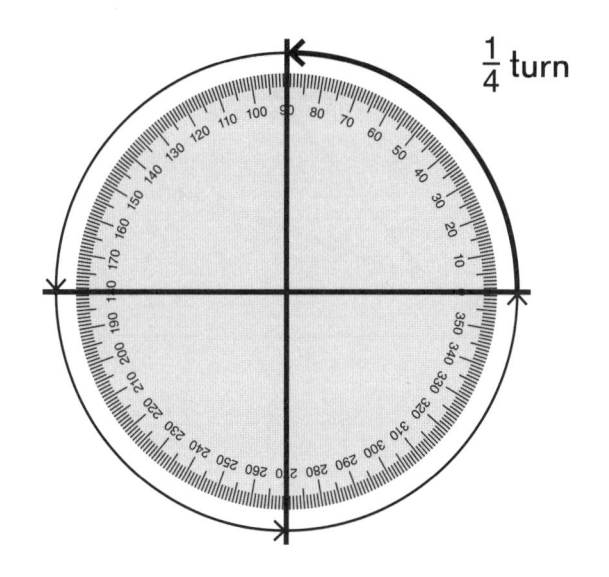

1 turn = 360°

$\frac{1}{4}$ turn = $\frac{1}{4}$ × 360° = ☐°

$\frac{1}{2}$ turn = $\frac{1}{2}$ × 360° = ☐°

$\frac{3}{4}$ turn = $\frac{3}{4}$ × 360° = ☐°

$\frac{1}{4}$ turn

2 Write 0°, 90°, 180°, or 360° in each blank to complete the table.

	acute angle	less than _____
	right angle	equal to _____
	obtuse angle	between _____ and _____
	straight angle	equal to _____
	reflex angle	between _____ and _____
	full turn	equal to _____

15-1 The Size of Angles

Practice

3 Use the right angle on a set square or the corner of an index card to identify each angle as an acute angle, a right angle, or an obtuse angle.

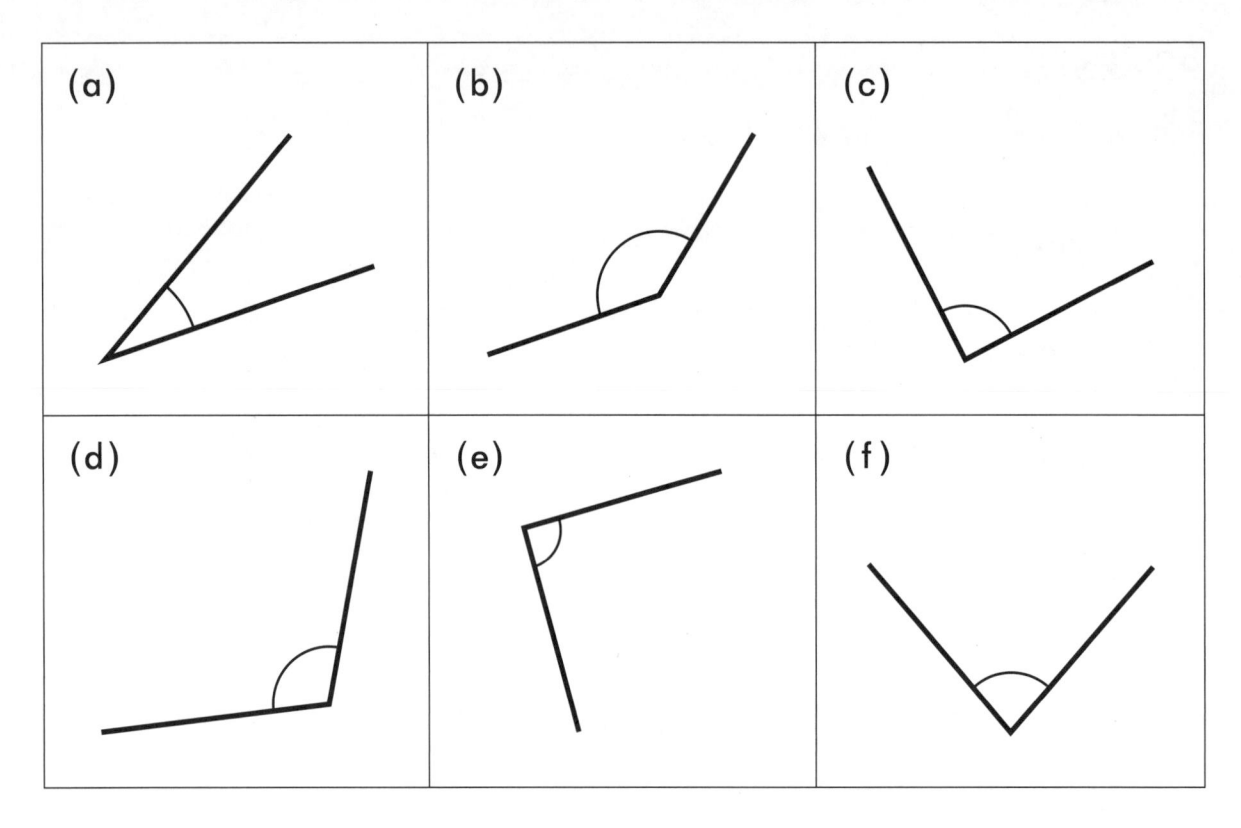

(a)

(b)

(c)

(d)

(e)

(f)

4 (a) A right angle is divided into 3 equal angles. What is the measure of each smaller angle in degrees?

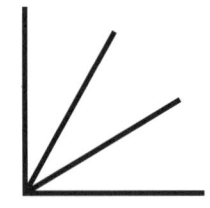

(b) A straight line is divided into 5 equal angles. What is the measure of each smaller angle in degrees?

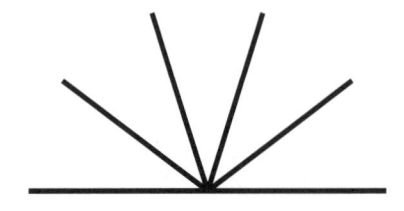

5) The fraction of a whole turn is given for each of these angles. Give the measure of each in degrees.

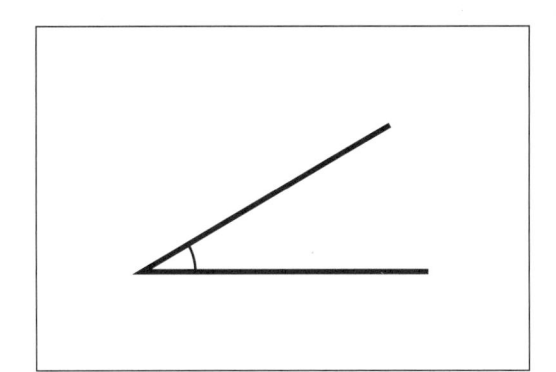

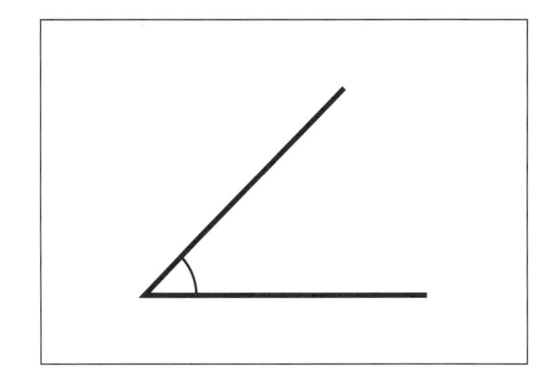

(a) $\frac{1}{12}$ turn = $\frac{1}{12}$ × 360° = ☐°

(b) $\frac{1}{8}$ turn = $\frac{1}{8}$ × 360° = ☐°

(c) $\frac{1}{6}$ turn = ☐°

(d) $\frac{1}{5}$ turn = ☐°

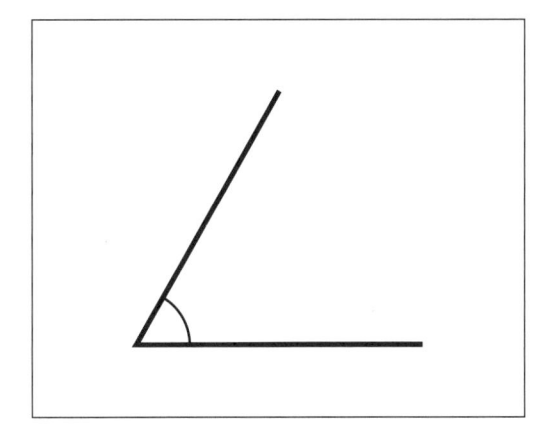

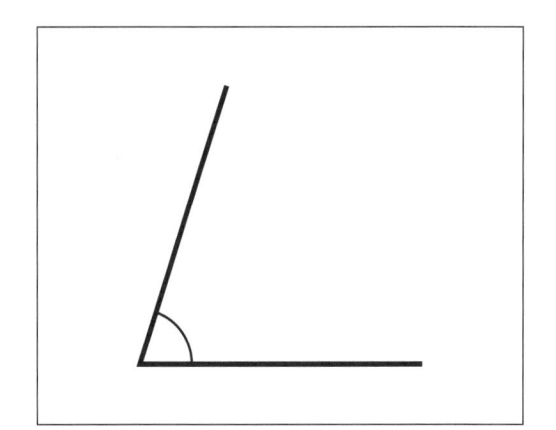

(e) $\frac{2}{3}$ turn = ☐°

(f) $\frac{4}{5}$ turn = ☐°

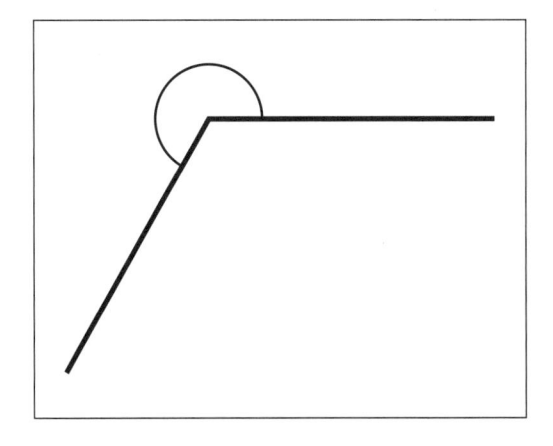

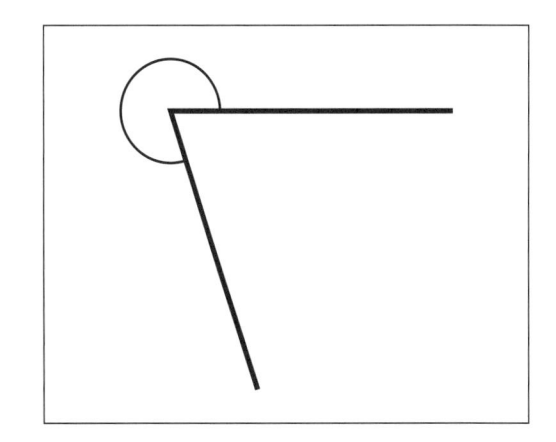

15-1 The Size of Angles

Challenge

6 The compass divides a circle into 8 equal angles. The directions on a compass going clockwise from the north (N) are, northeast (NE), east (E), southeast (SE), south (S), southwest (SW), west (W), and northwest (NW).

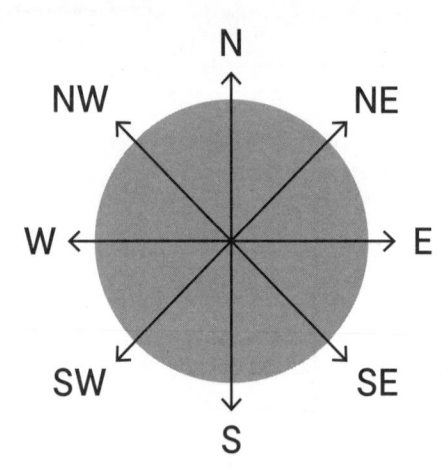

(a) Connor is facing south. If he turns 225° counterclockwise, what direction will he face?

(b) Rodrigo is facing northeast. If he turns 135° clockwise, what direction will he face?

(c) Ximena is facing southwest. If she makes a $\frac{1}{4}$ turn to her right and then a $\frac{3}{4}$ turn to her left, what direction will she be facing?

(d) Sara is facing southeast. If she turns 45° clockwise and then 315° counterclockwise, what direction will she be facing?

Exercise 2

Basics

1 What is the size of each angle in degrees?

(a) ∠a = ☐°

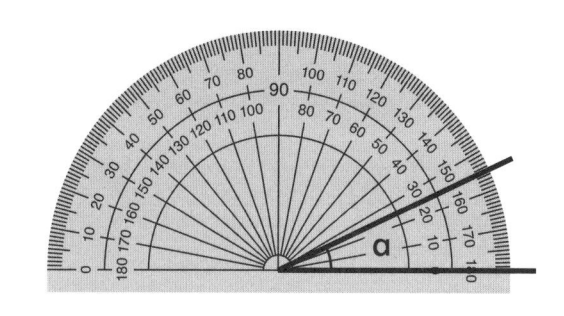

(b) ∠b = ☐°

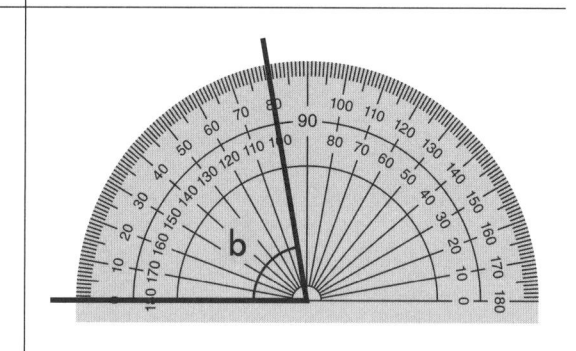

(c) ∠c = ☐°

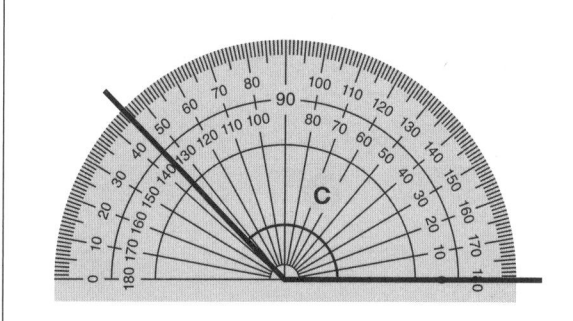

(d) ∠d = ☐°

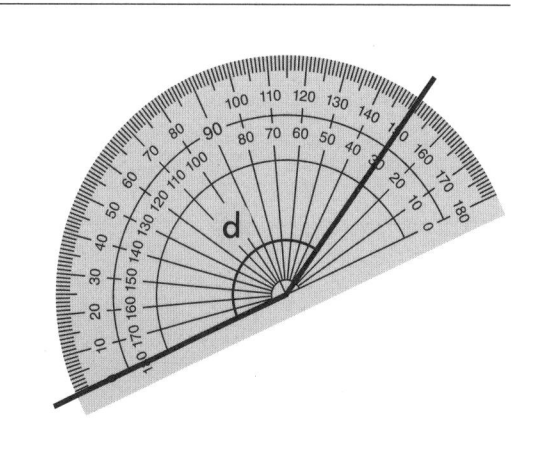

2 Match, without measuring.

30°

45°

60°

90°

135°

180°

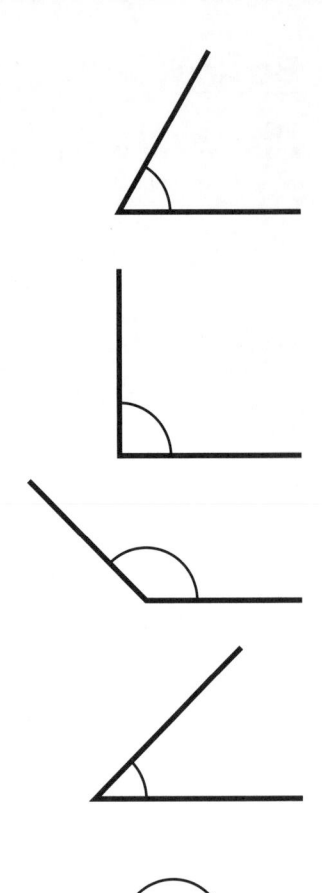

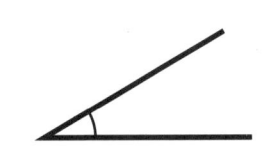

3 In the diagrams below, name the marked angles using the given letters.

(a) ∠ [　　　]

(b) ∠ [　　　]

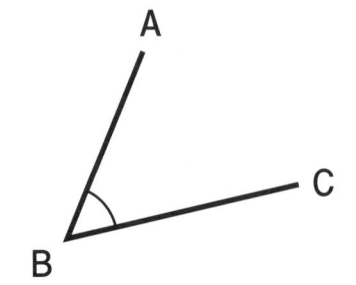

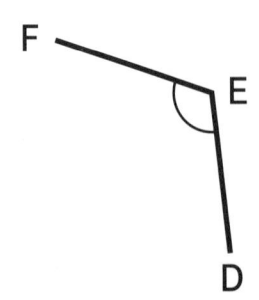

Practice

4) Circle the angle that is 75°. Use estimation.

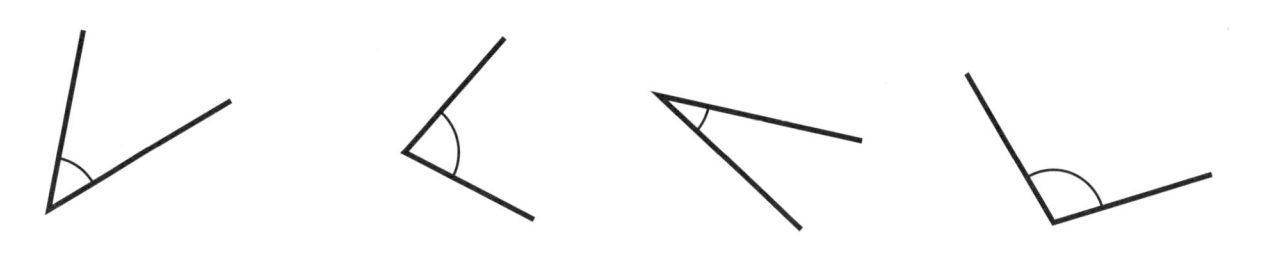

5) Estimate the size of each angle. Then measure each angle with a protractor.

∠ABC ≈	∠DEF ≈
∠ABC =	∠DEF =

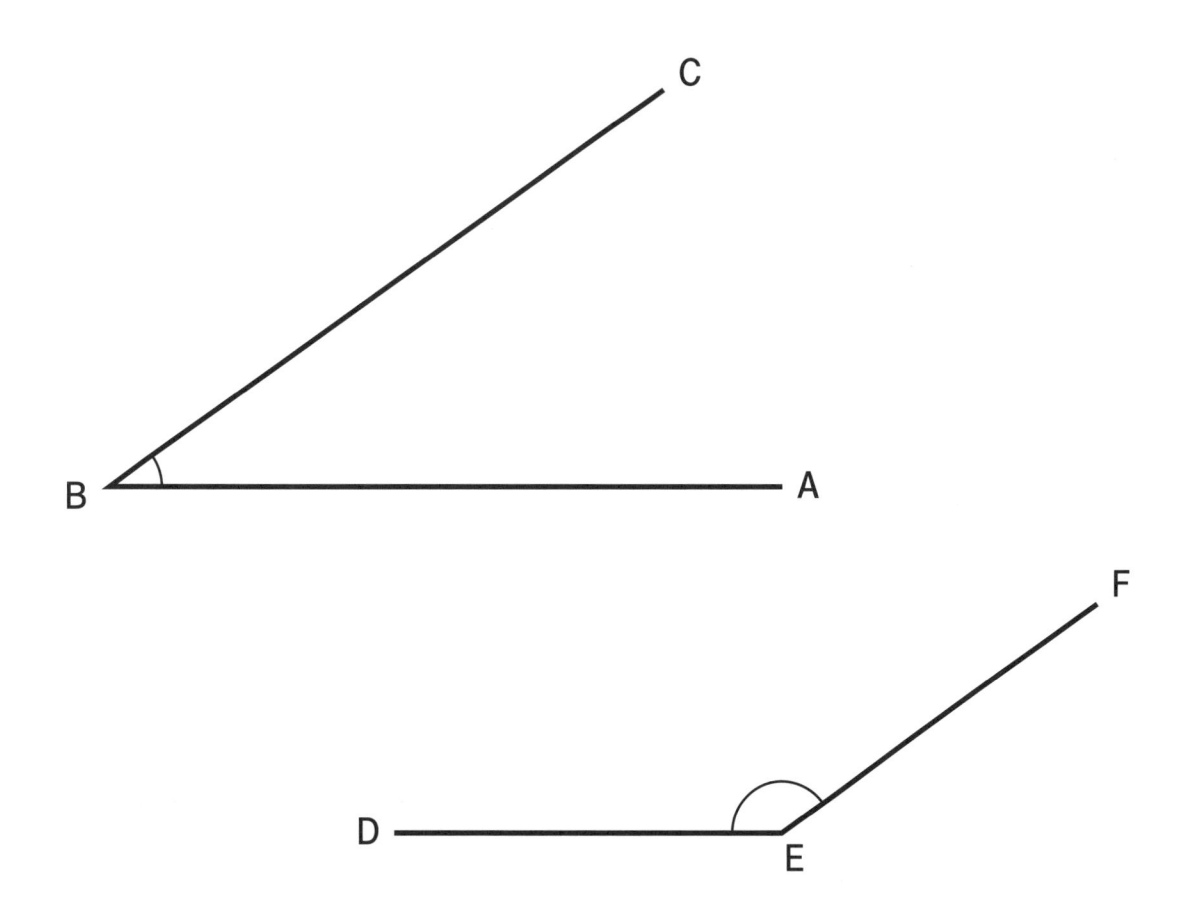

∠GHI ≈	∠JKL ≈	∠MNO ≈	∠PQR ≈
∠GHI =	∠JKL =	∠MNO =	∠PQR =

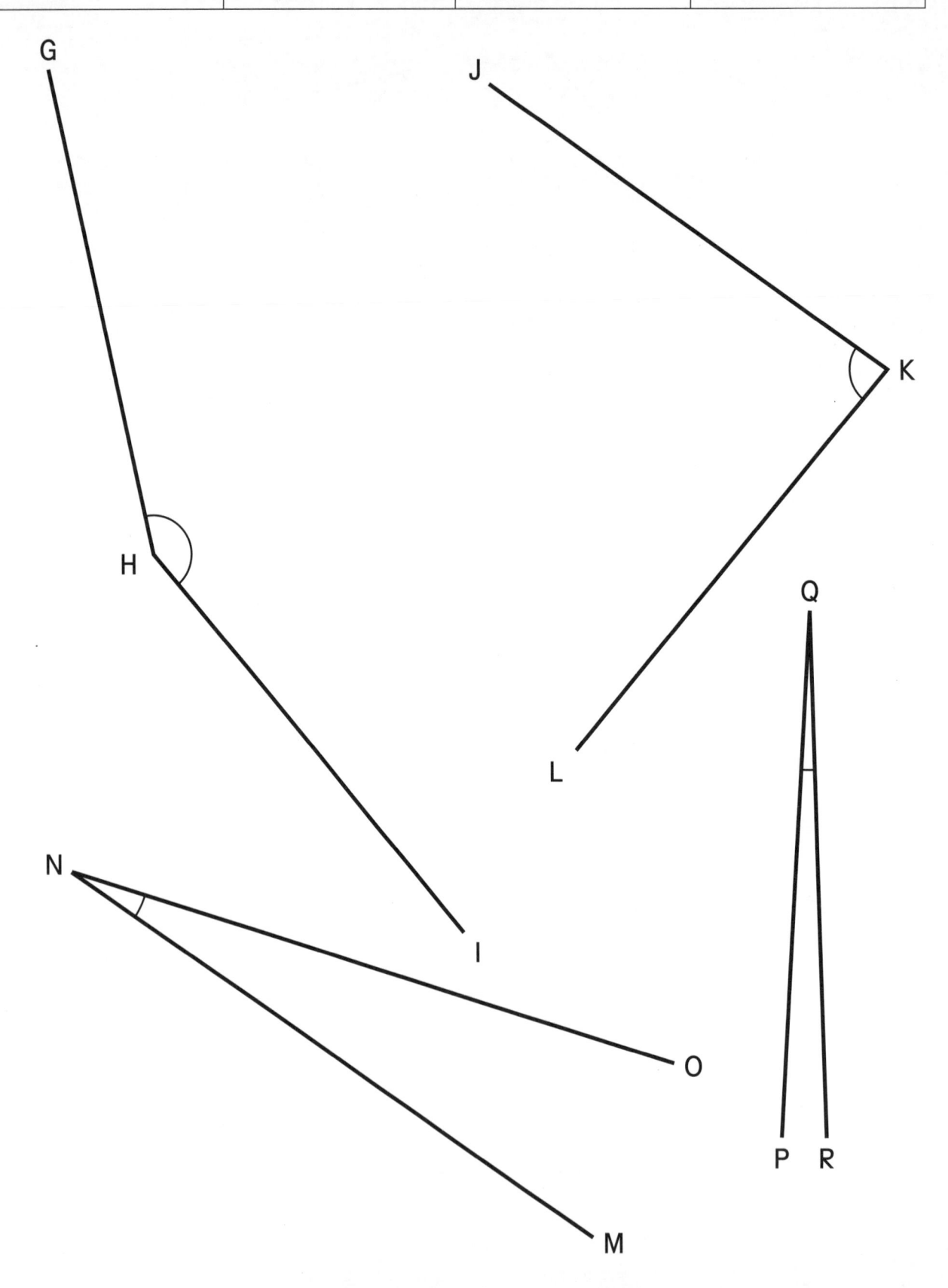

15-2 Measuring Angles

Exercise 3

Basics

1 Complete the drawing of each angle.

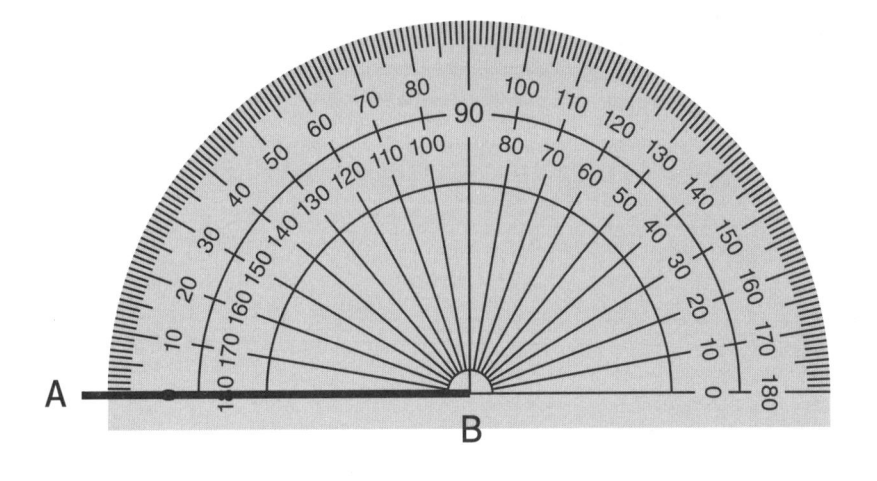

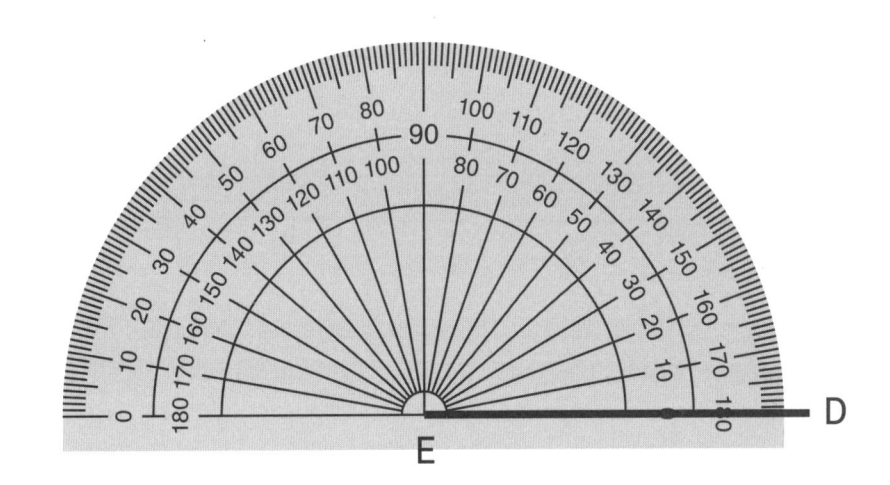

(a) ∠ABC = 45°

(b) ∠DEF = 62°

(c) ∠GHI = 118°

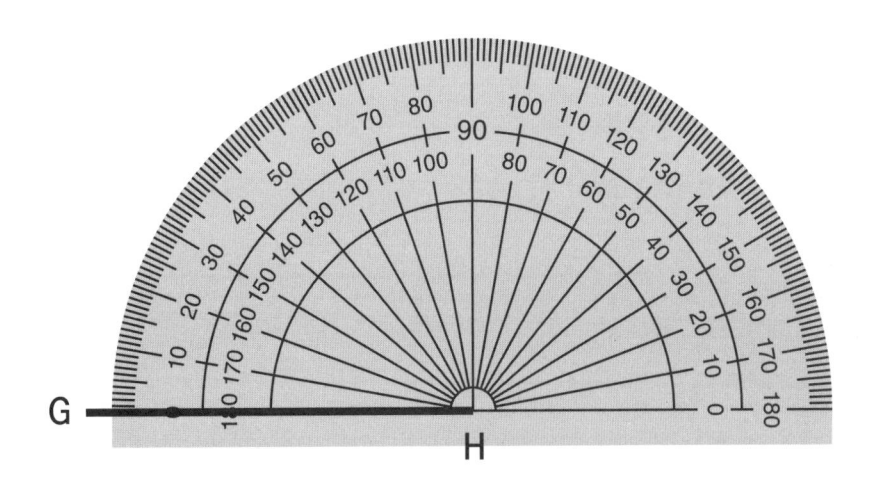

15-3 Drawing Angles 139

Practice

2 Draw a line through the correct point to get the required angle. (Use a protractor to help you choose the correct point.) Mark the angle.

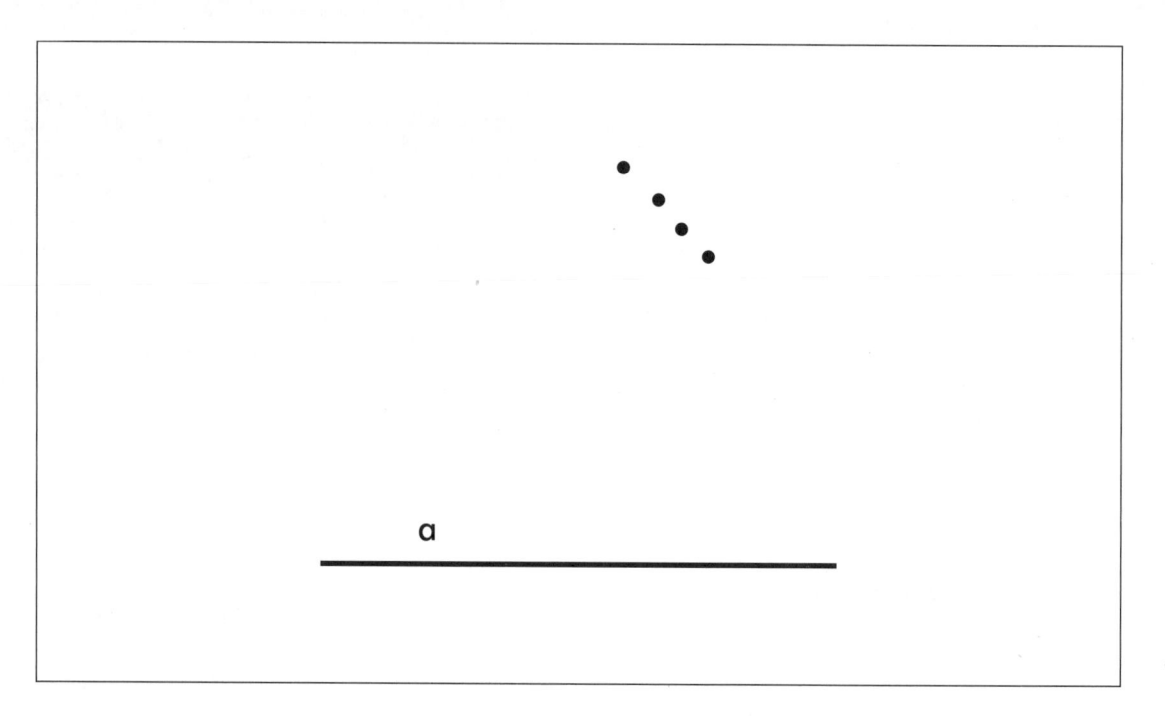

(a) ∠a = 42°

(b) ∠b = 157°

140 15-3 Drawing Angles

(c) ∠c = 75°

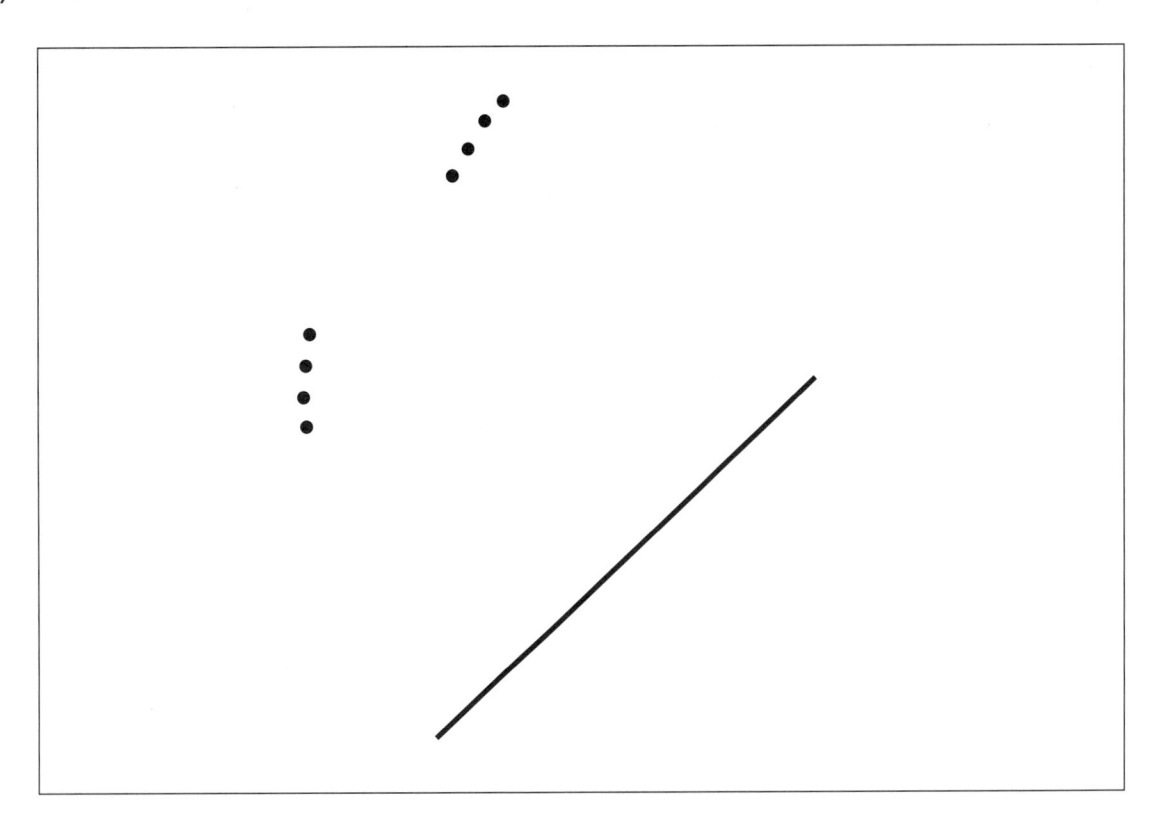

(d) ∠d = 103°

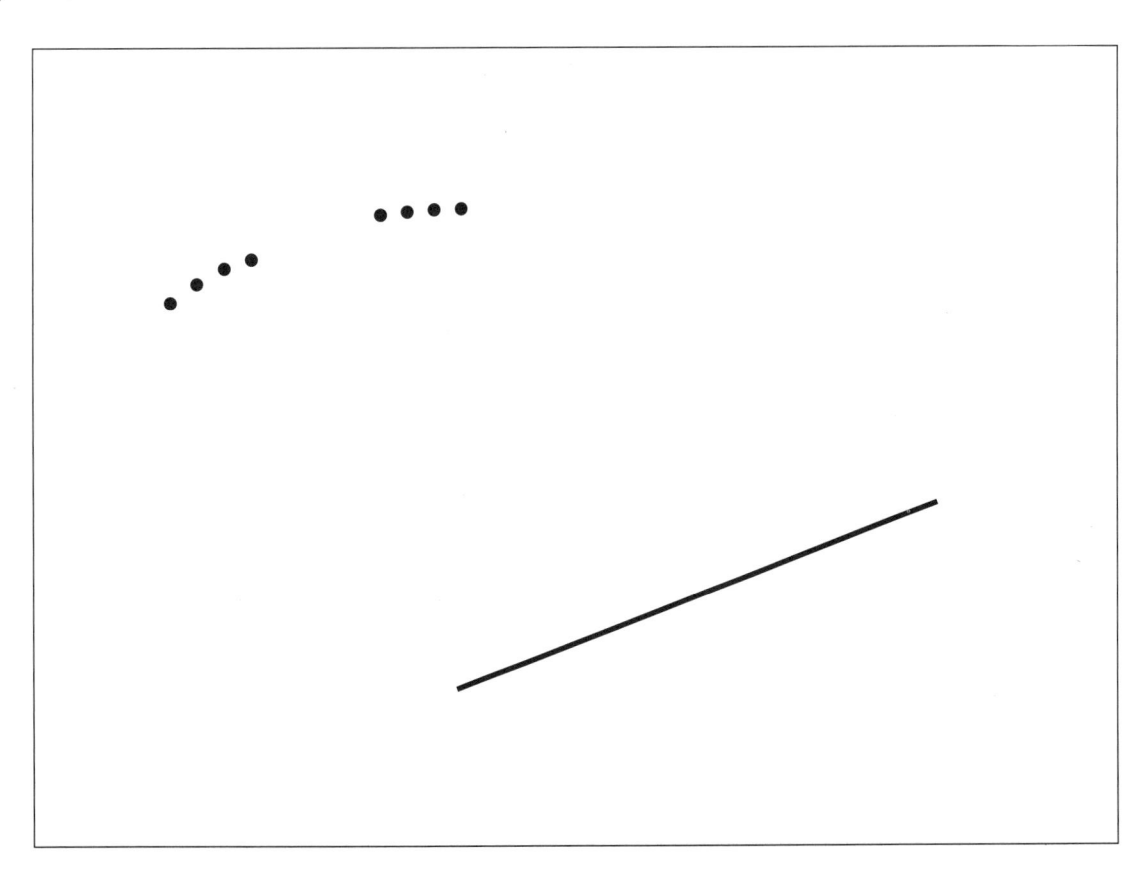

15-3 Drawing Angles

3 Draw a 25° angle.

4 Draw a 135° angle.

5 Draw a quadrilateral with 4 angles that are each 90°.

Challenge

6 Draw a quadrilateral with 3 angles the same size, but not 90°. All the inside angles should be less than 180°.

Exercise 4

Basics

1 ∠ABC = 35° and ∠CBD = 45°. What is the measure of ∠ABD?

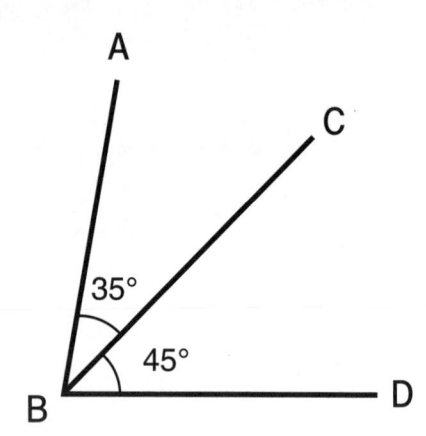

∠ABC + ∠CBD = ☐° + ☐°

= ☐°

∠ABD = ☐°

2 ∠EFG is a right angle. Find the measure of the unknown angle.

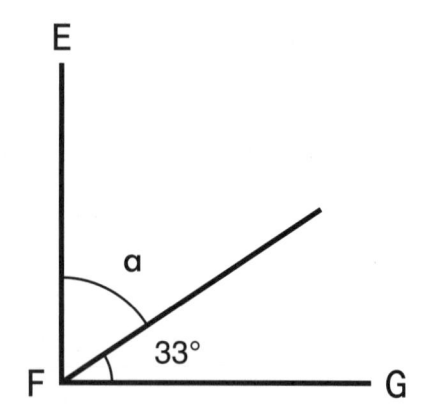

90° − 33° = ☐°

∠a = ☐°

Practice

3 JK is a straight line. Find the measure of the unknown angle.

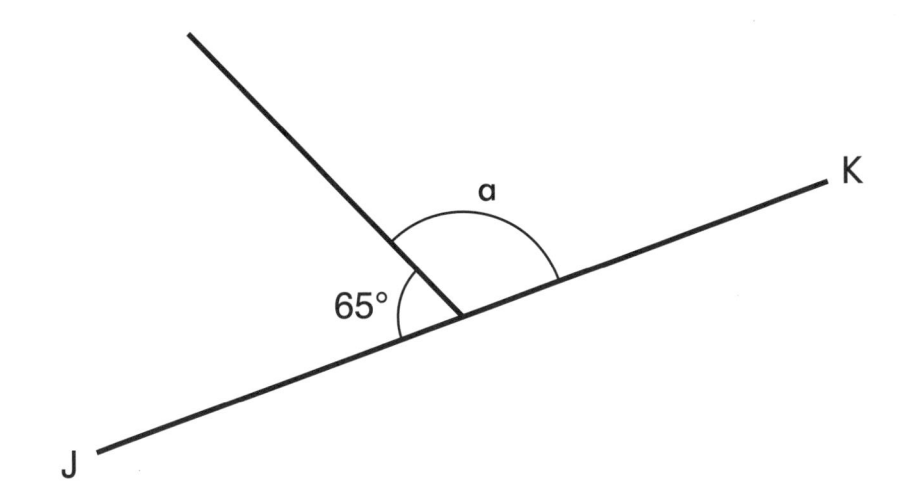

4 Calculate to find the measure of the unknown marked angles in each rectangle.

| ∠a = | ∠b = | ∠c = | ∠d = |

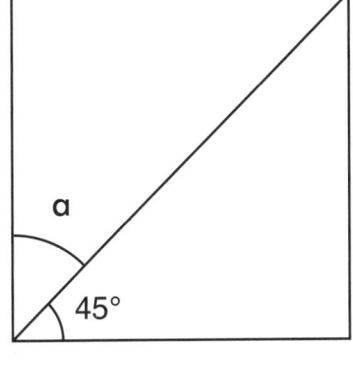

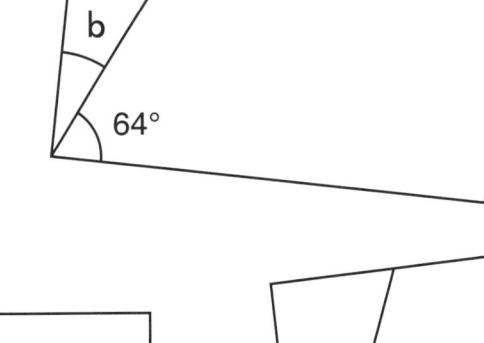

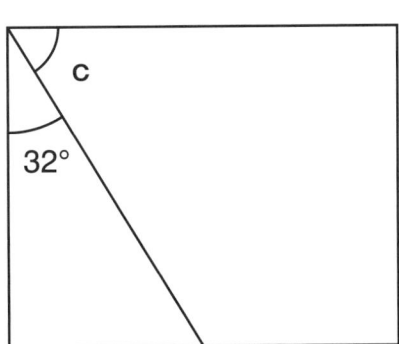

15-4 Adding and Subtracting Angles

Challenge

5 AC is a straight line. Find the measure of ∠BED.

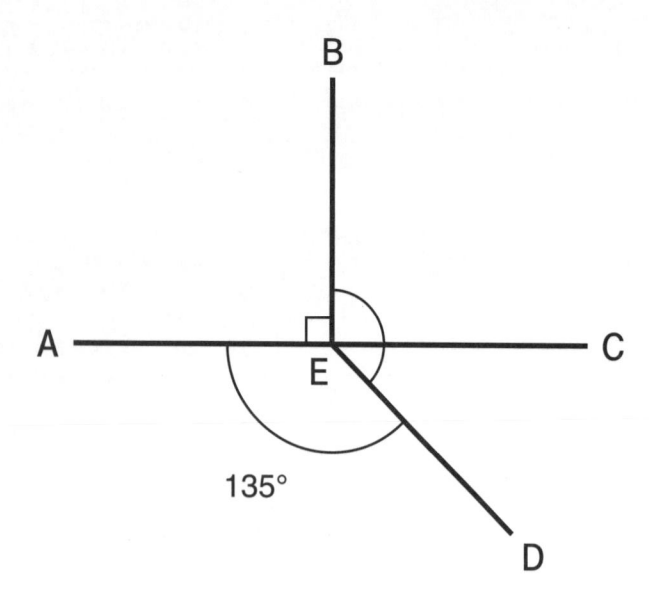

6 AB is a straight line. ∠f is 4 times as large as ∠g. Find the measure of ∠f.

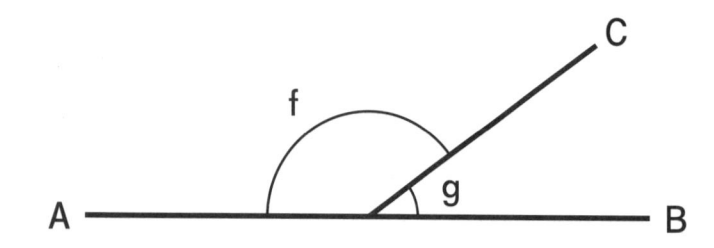

Exercise 5

Basics

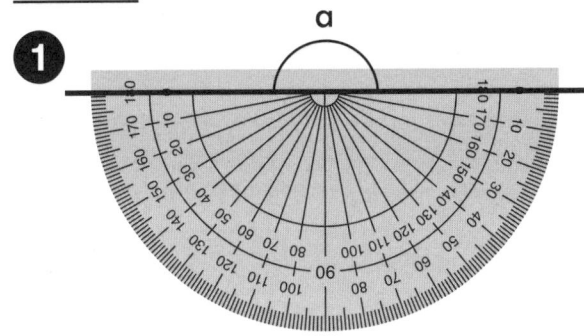

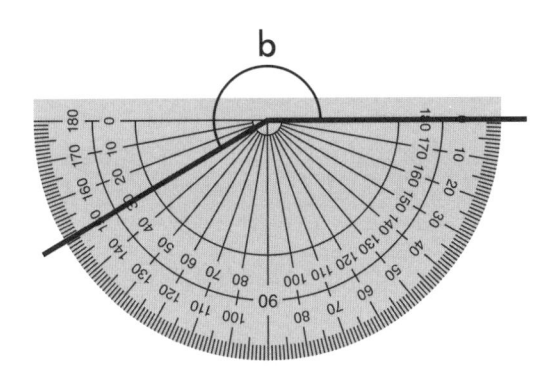

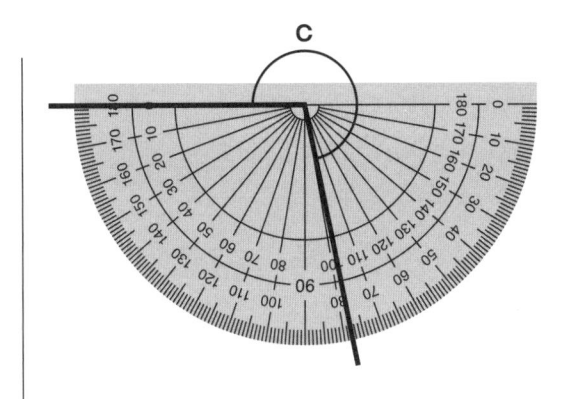

1

∠a = 180°

∠b = 180° + ☐°

  = ☐°

∠c = 180° + ☐°

  = ☐°

2 1 full circle = 360°

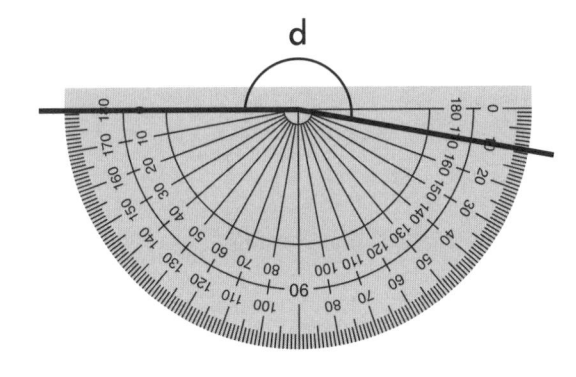

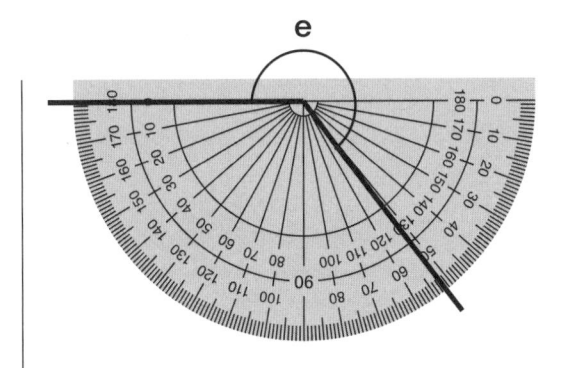

∠d = 360° − ☐°

  = ☐°

∠e = 360° − ☐°

  = ☐°

15-5 Reflex Angles 147

Practice

3 Estimate the size of each reflex angle. Then measure each angle with a protractor.

∠a ≈	∠b ≈	∠c ≈
∠a =	∠b =	∠c =

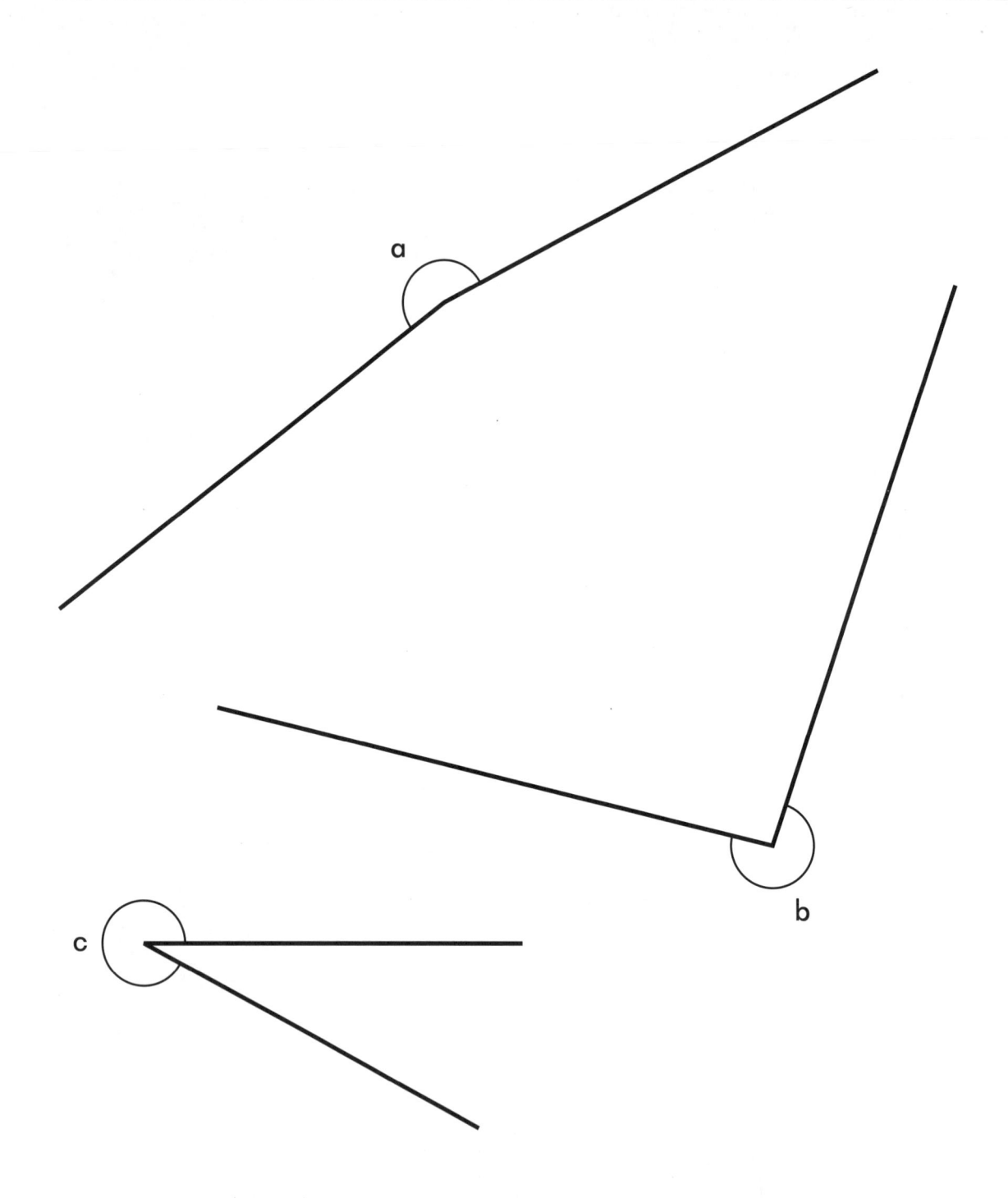

4 Draw a 200° angle.

5 Draw a 317° angle.

6 What is the measure of each exterior angle of a rectangle?

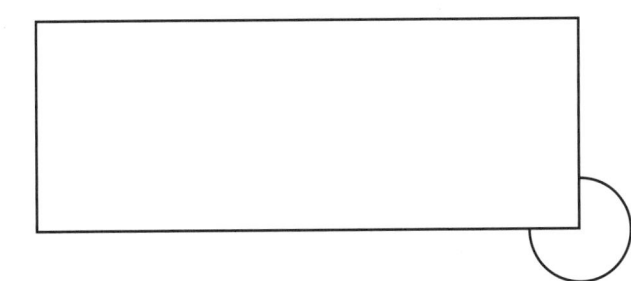

15-5 Reflex Angles

Challenge

7 Find the measure of ∠a.

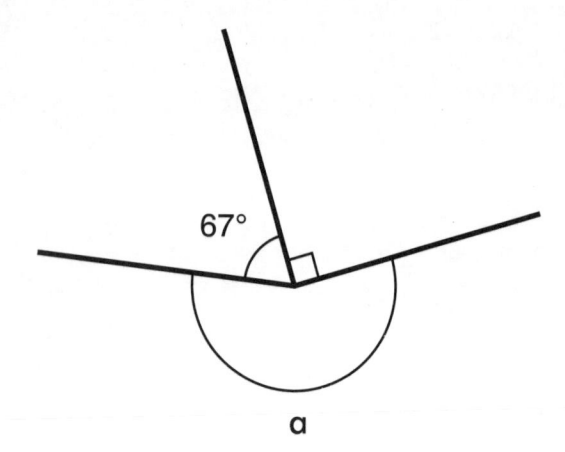

8 ∠b is three times as large as ∠a. ∠c is twice as large as ∠b. What is the measure of ∠c?

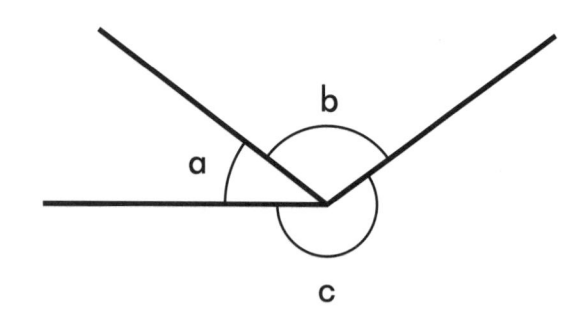

Exercise 6

Check

1 Write the type of angle for each internal angle in the shapes below (acute, right, obtuse, reflex). Then measure each labeled angle. Write each measure.

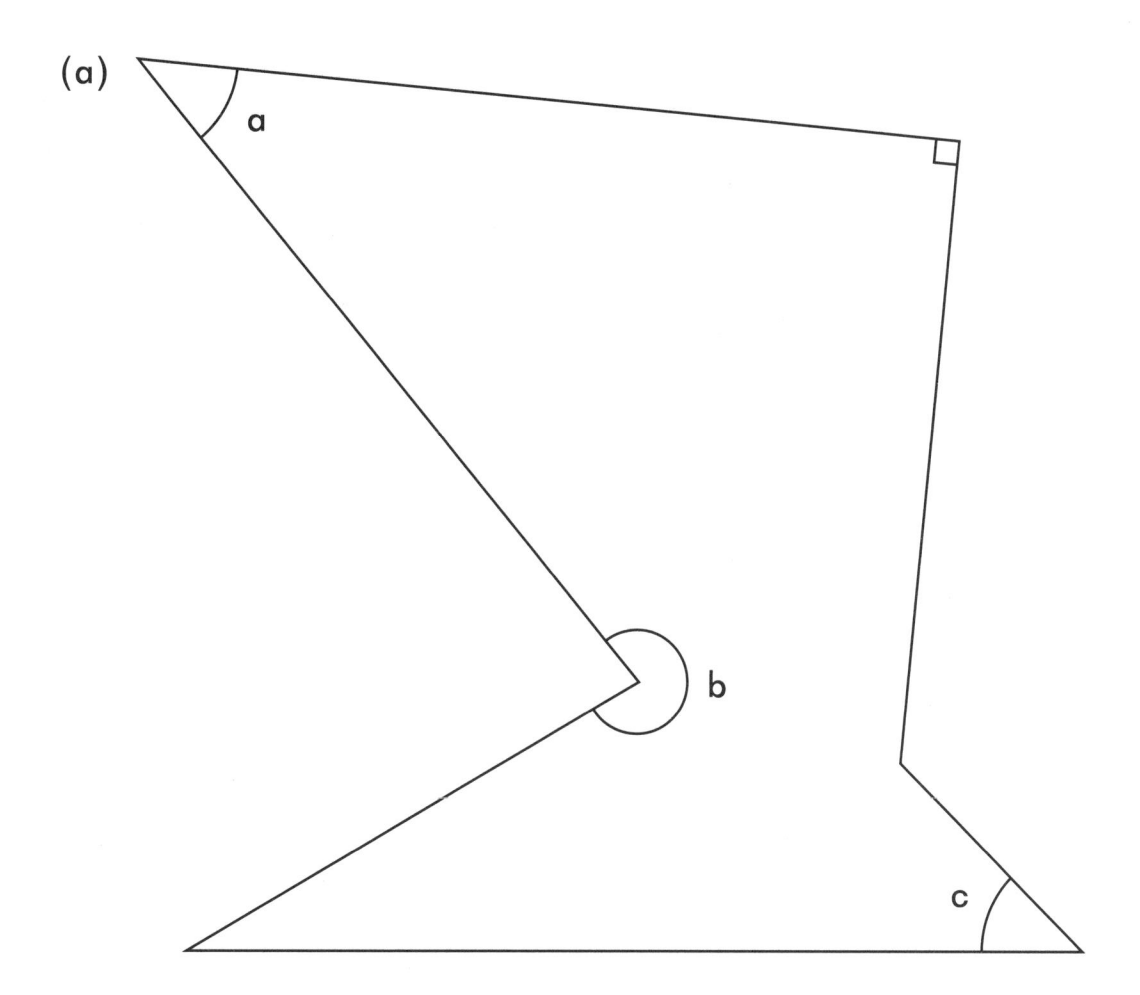

(a)

(b)

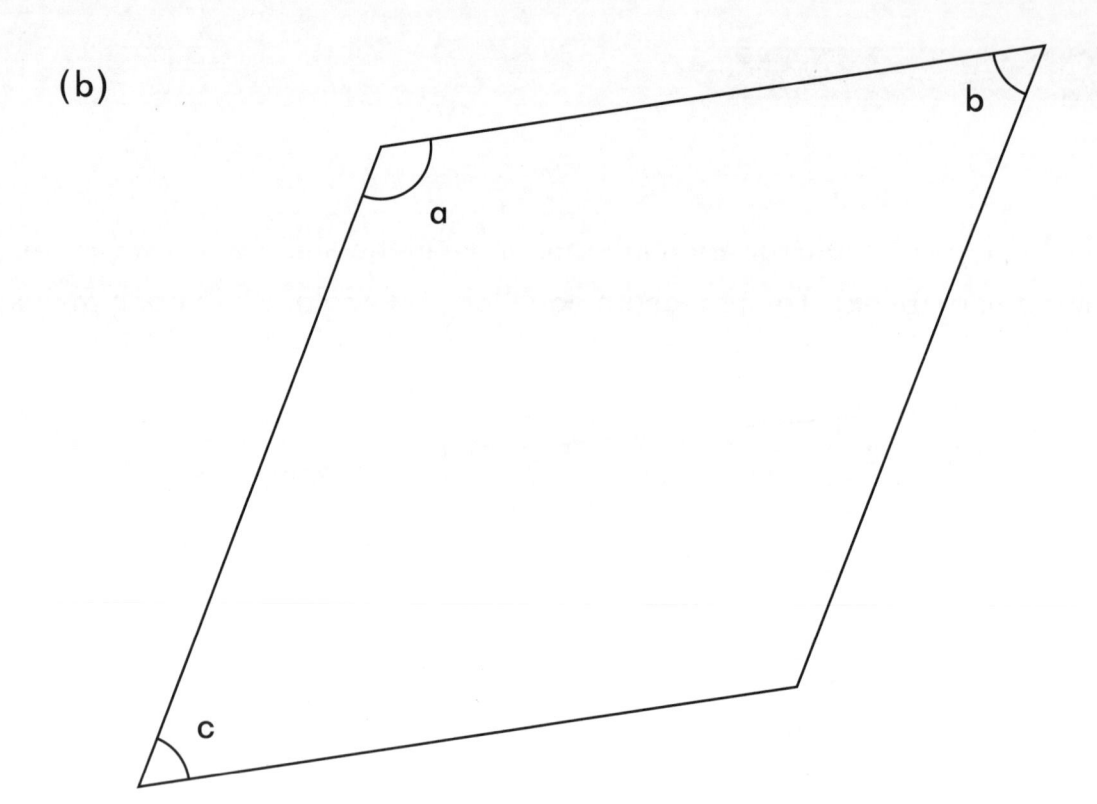

[quadrilateral with angles labeled a, b, c]

2 (a) Draw an angle that measures 74°.

(b) Calculate the measure of the reflex angle that was formed in your drawing for (a).

3 (a) Draw an angle that measures 346°.

(b) Calculate the measure of the acute angle that was formed in your drawing for (a).

4 ABCD is a rectangle. Find the measure of the Angles ADE and DEC.

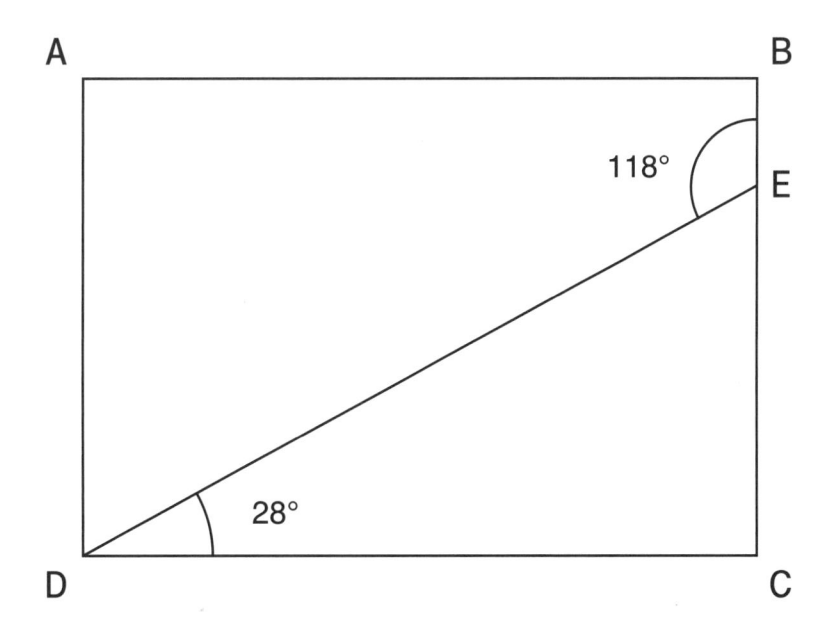

∠ADE =

∠DEC =

15-6 Practice

5 WX and YZ are straight lines. Find the measure of Angles WPZ, WPY, and YPX.

| ∠WPZ = | ∠WPY = | ∠YPX = |

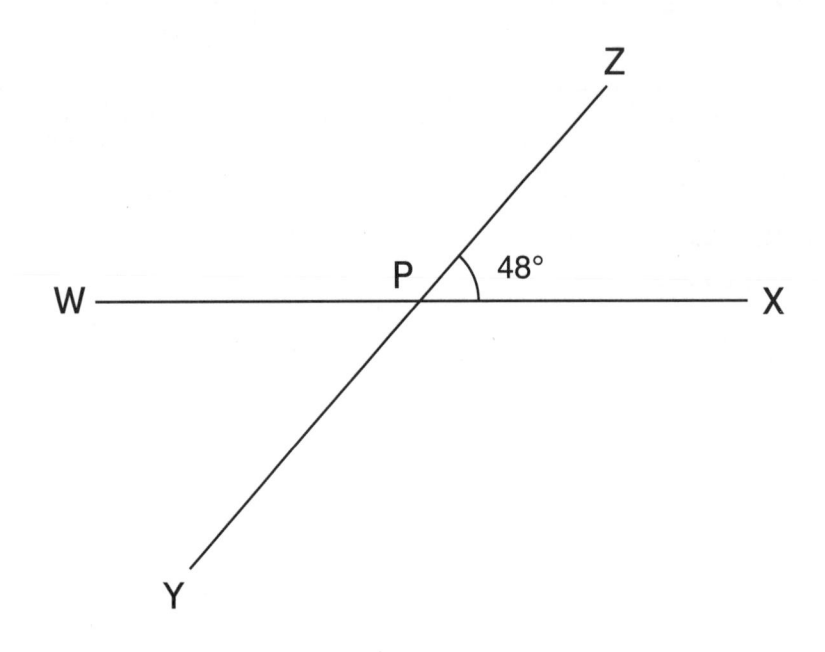

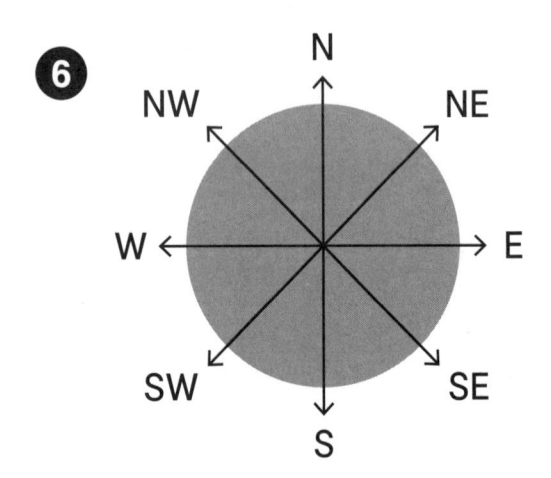

6

Sara is facing west. She turns 270° counterclockwise. What direction is she now facing?

Challenge

7 How many right angles does the minute hand move through between 12:00 a.m. and 3:00 a.m.?

8 (a) How many degrees does the hour hand turn each hour?

(b) How many degrees does the minute hand turn each minute?

9 What is the measure of the obtuse angle between the hands at 4:00 p.m.?

10 What is the measure of the reflex angle between the hands at 10:30 p.m.?

Chapter 16 Lines and Shapes

Exercise 1

Basics

1 The two lines are perpendicular to each other because they intersect at _____ angles.

AB ⊥ CD

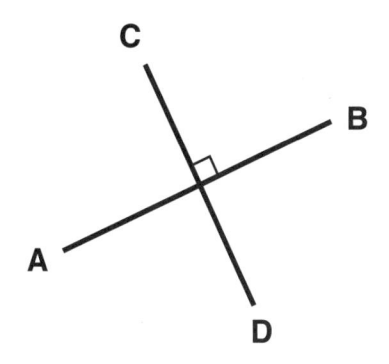

2 Write a check mark in the box if the lines are perpendicular to each other. Extend the lines to intersect if needed. Use a set square to identify right angles.

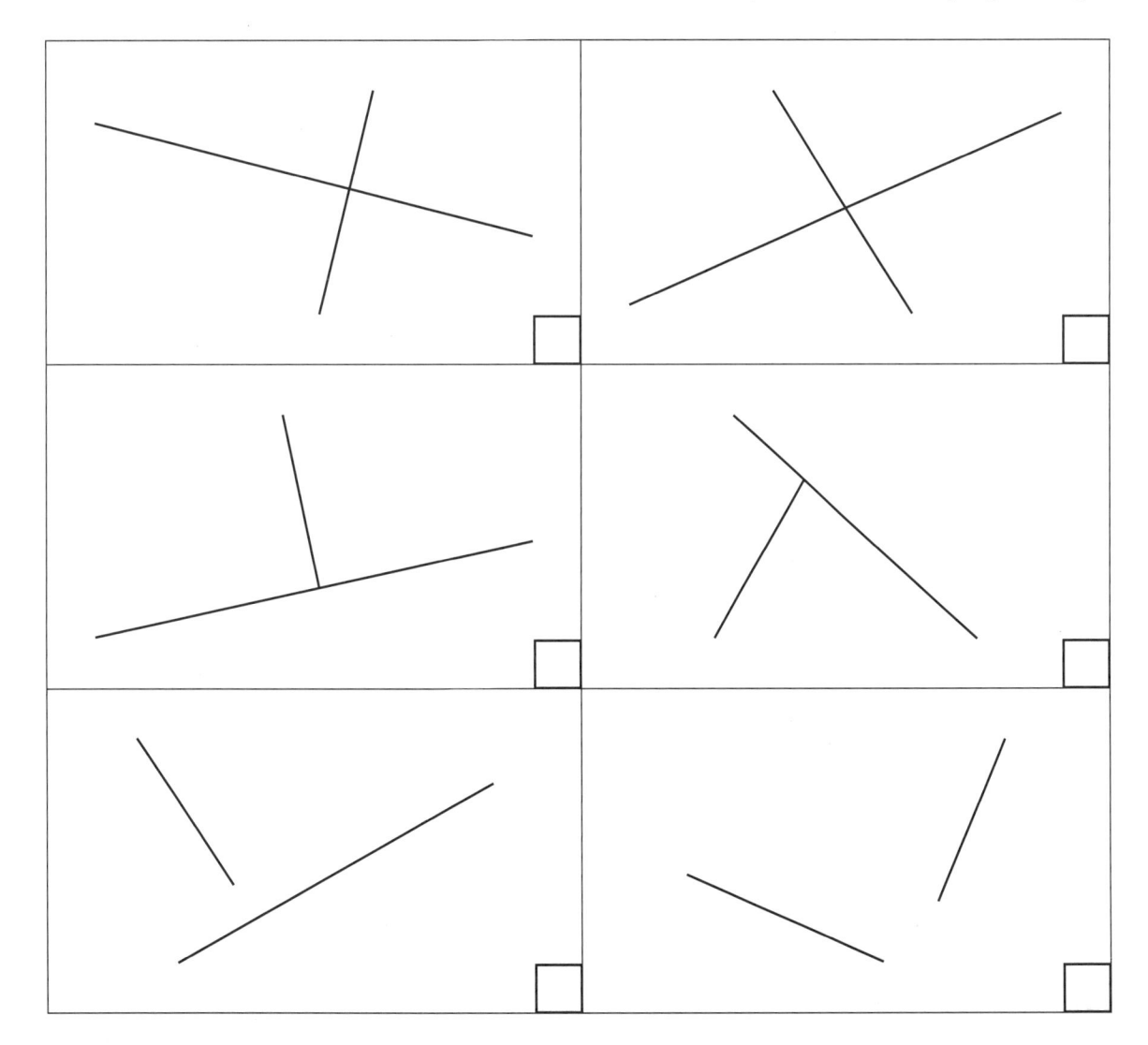

16-1 Perpendicular Lines

Practice

3 Name the pairs of perpendicular sides in Rectangle ABCD. The first has been done for you.

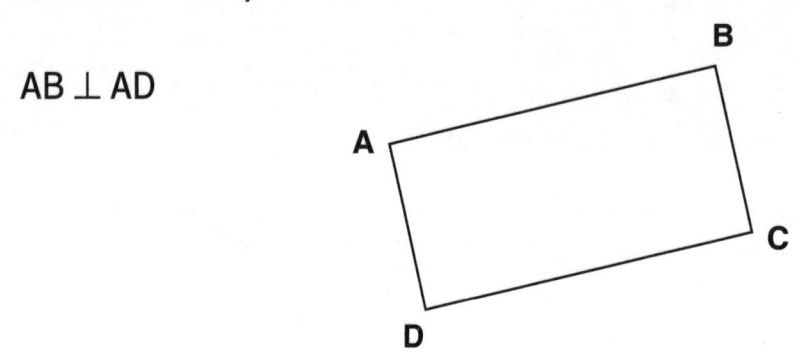

AB ⊥ AD

4 For each of the following figures, name each pair of perpendicular lines. Use a set square to identify right angles.

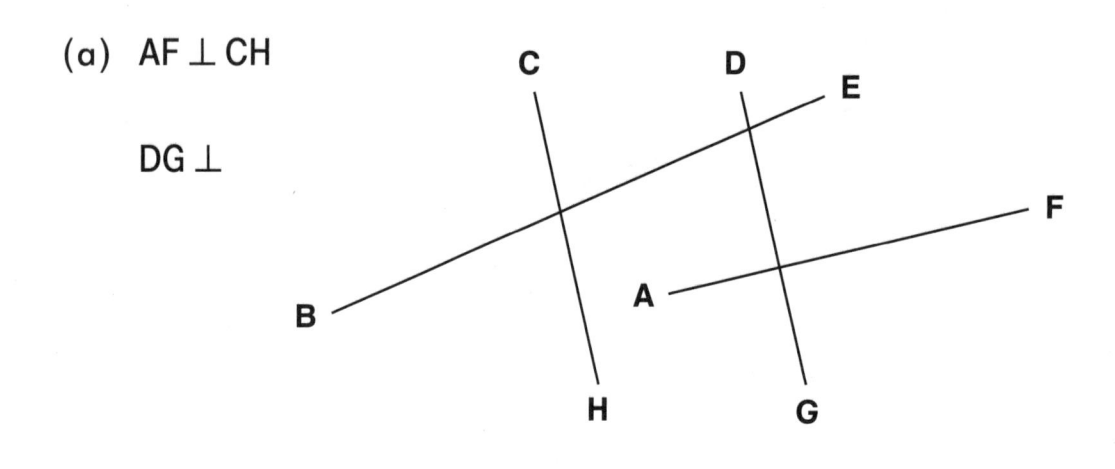

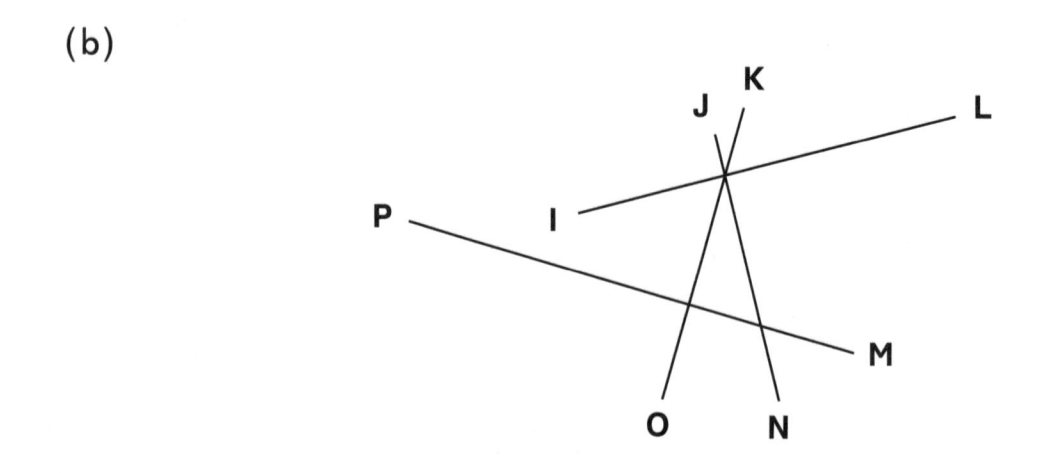

(a) AF ⊥ CH

DG ⊥

(b)

158 16-1 Perpendicular Lines

(c)

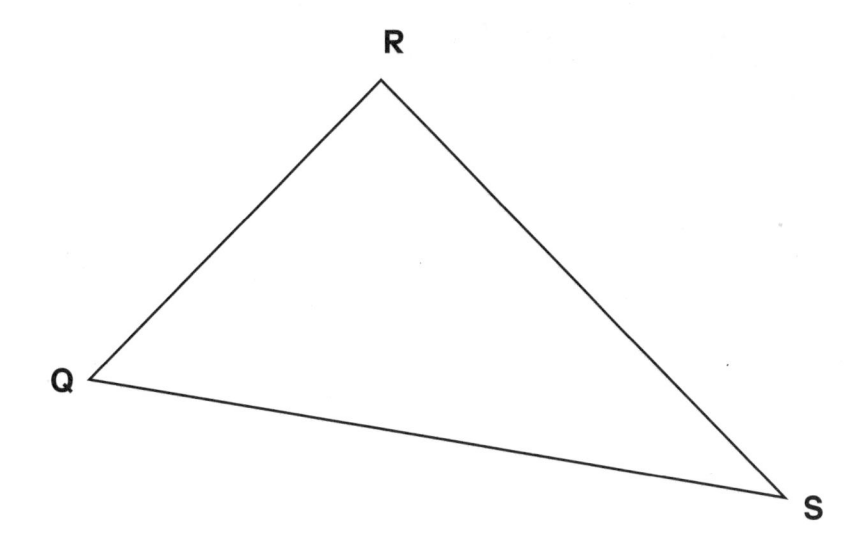

(d)

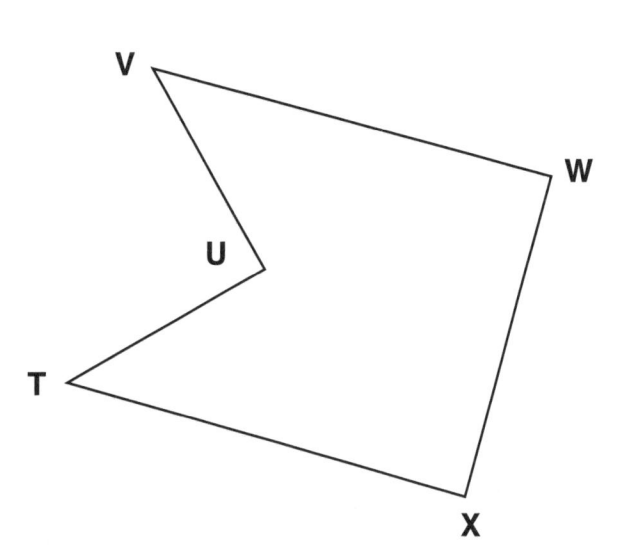

(e)

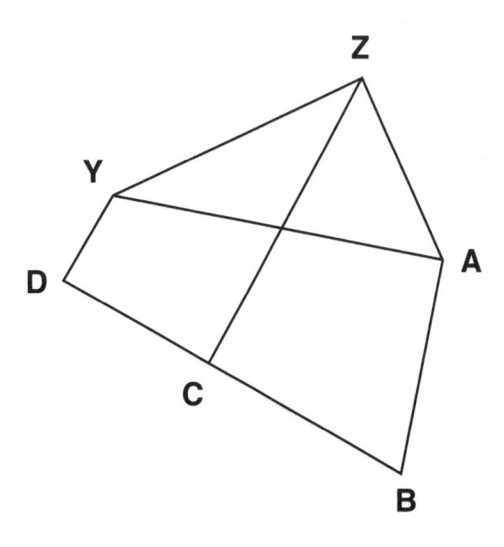

16-1 Perpendicular Lines

Exercise 2

Basics

1

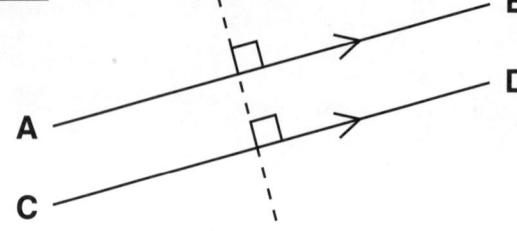

The two lines marked with arrowheads are parallel to each other because they are both _____ to a third line.

AB || CD

2 EFGH is a rectangle. Name the pairs of parallel sides.

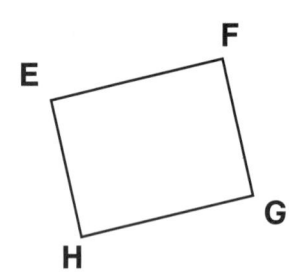

3 Name the pairs of parallel lines in the figure below. Use a set square to check if two lines are perpendicular to a third line.

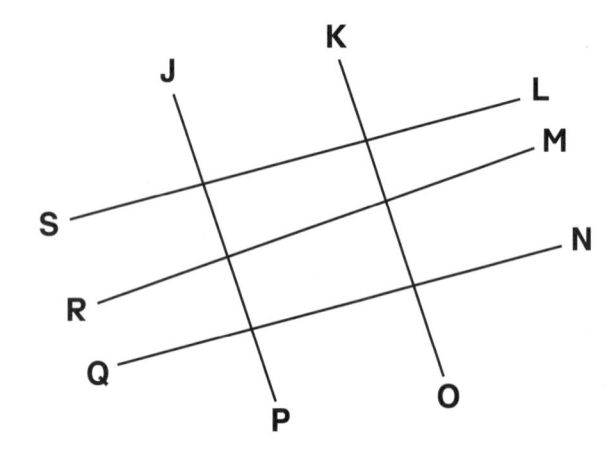

160   16-2 Parallel Lines

Practice

4 Name pairs of parallel lines in the following figures. Use a set square to identify right angles.

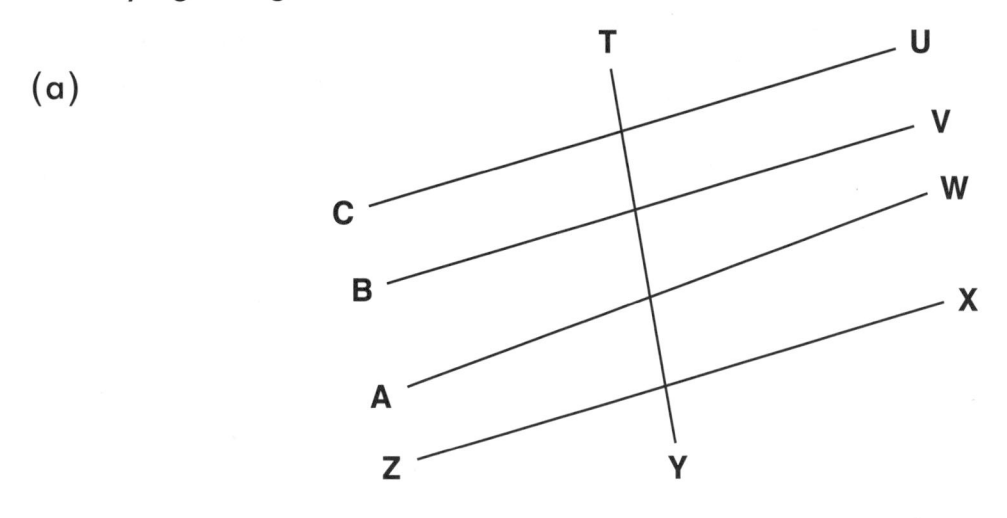

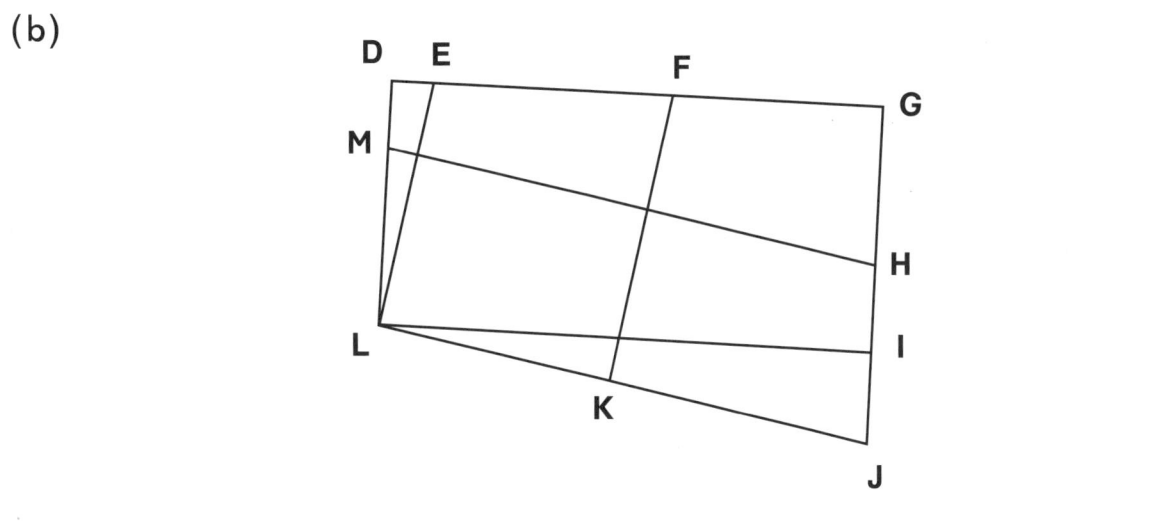

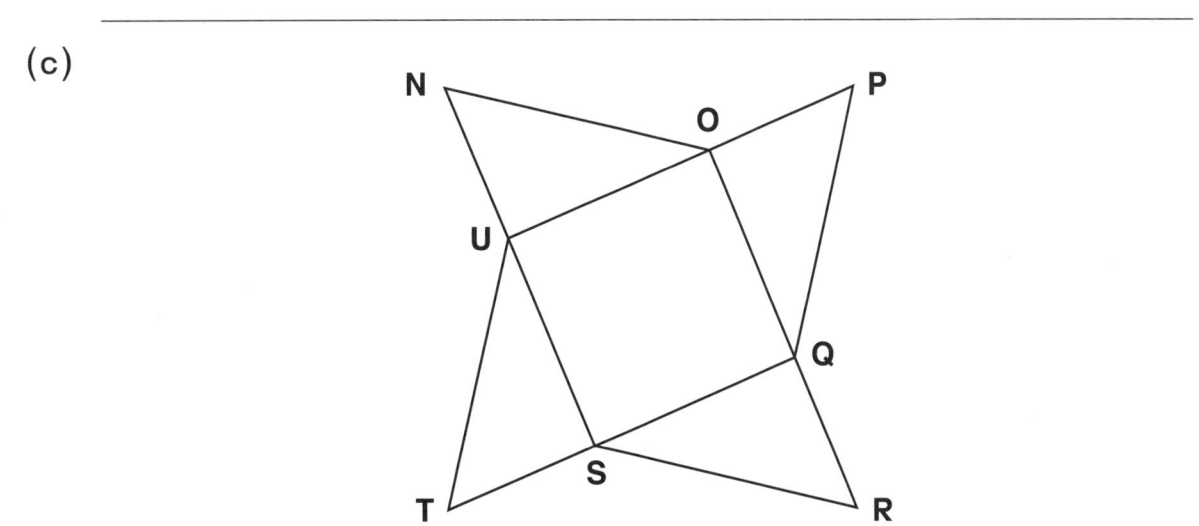

(a)

(b)

(c)

16-2 Parallel Lines

Exercise 3

Basics

1 (a) Use a set square. Draw a line through C perpendicular to AB. Draw a line through D perpendicular to AB.

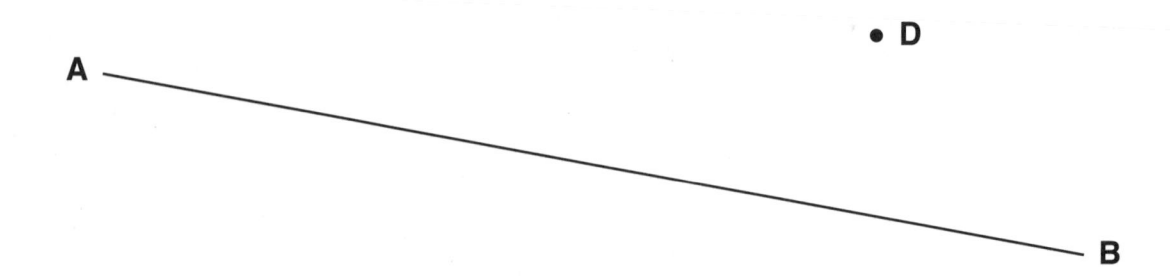

C •

• D

A ———————————————— B

(b) Are the two lines you drew parallel to each other?

2 Use a ruler and a set square to draw a line through G that is parallel to EF.

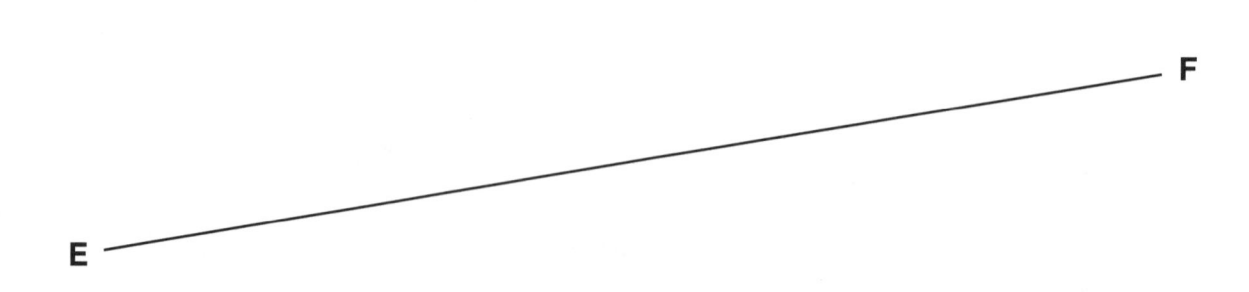

G •

E ———————————————— F

162　　16-3 Drawing Perpendicular and Parallel Lines

Practice

3 Use a ruler and a set square to draw one line perpendicular and one line parallel to the given lines through the given points.

(a)

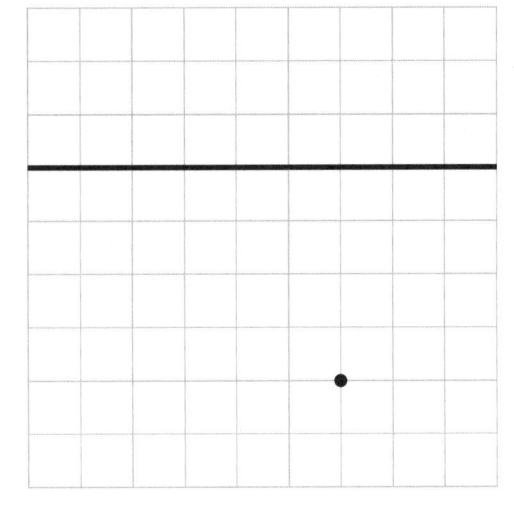

(b)

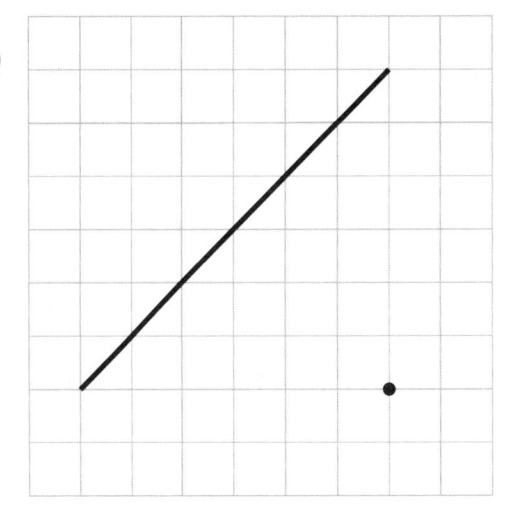

(c)

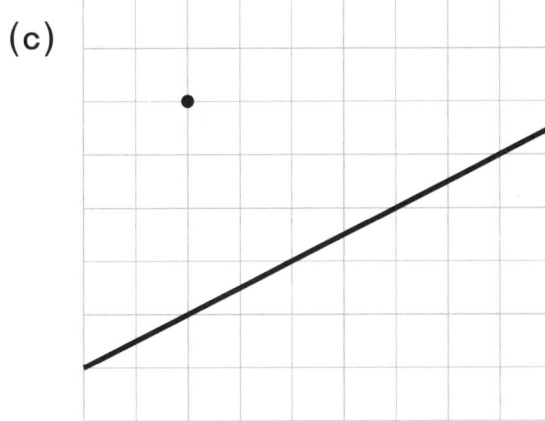

(d)

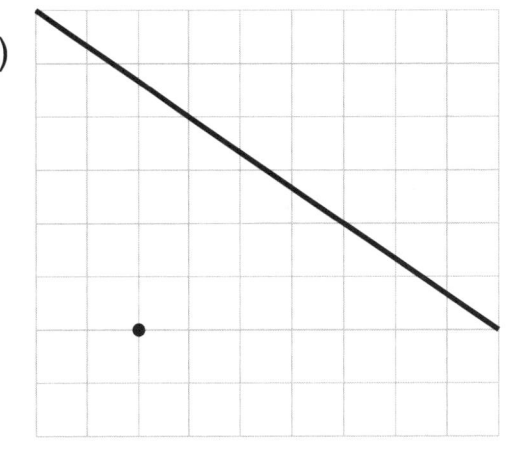

(e)

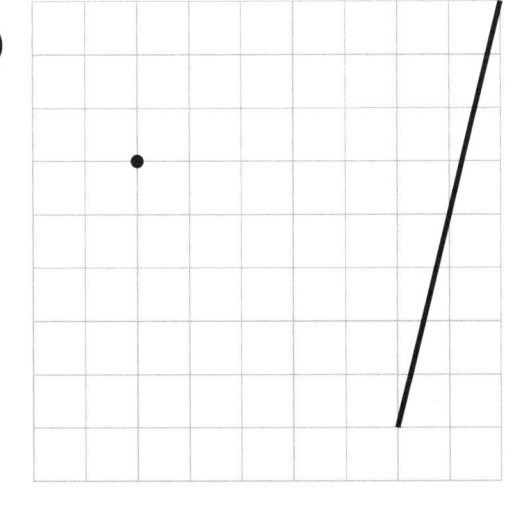

(f)

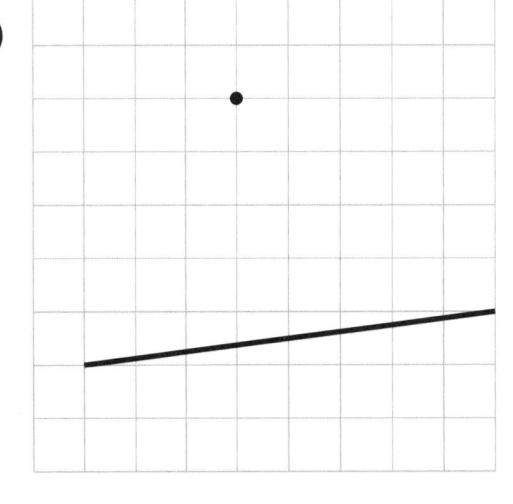

16-3 Drawing Perpendicular and Parallel Lines

4 Use a ruler and set square to identify parallel lines. Name all pairs of parallel lines.

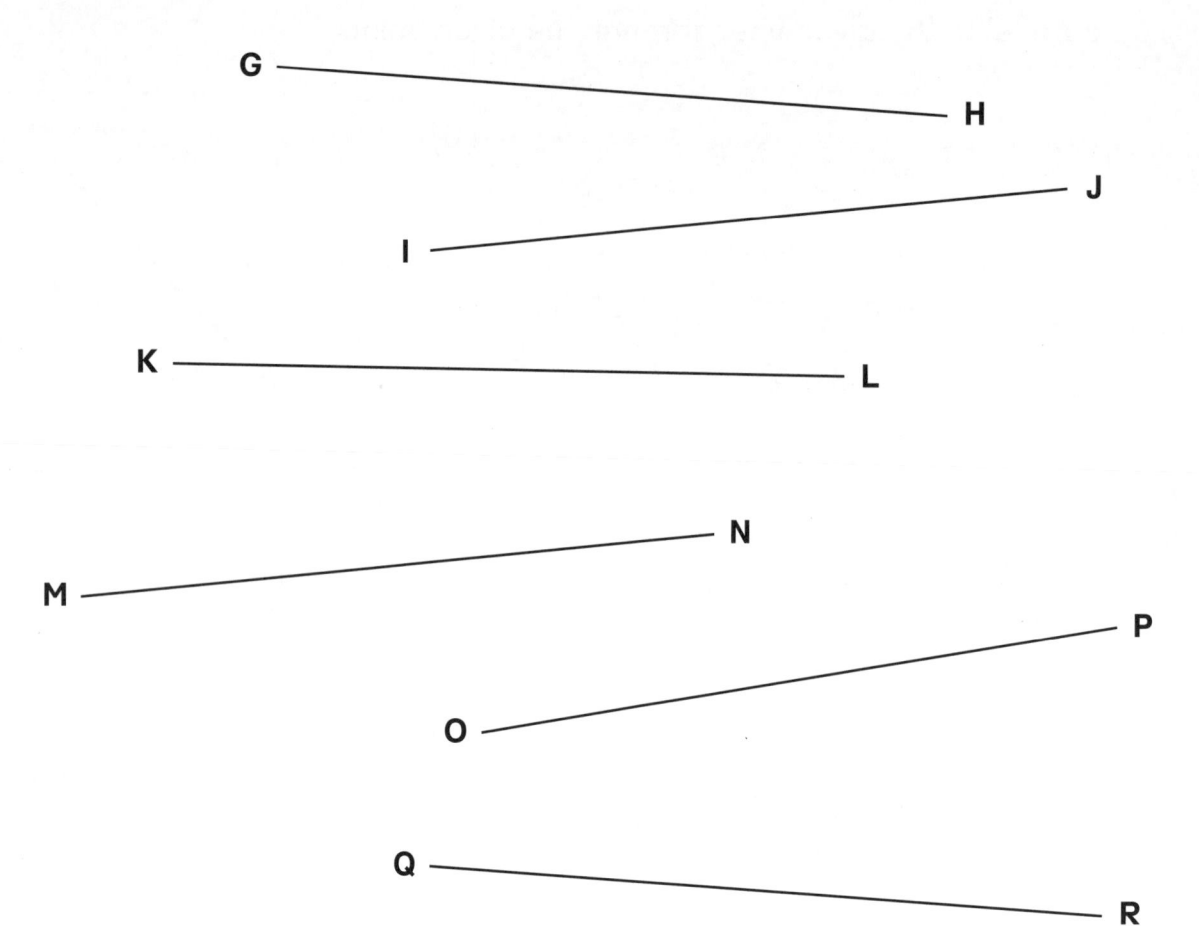

G ———————————
              ——— H

                  ——— J
    I ———————————

K ——————————————— L

M ———————————
         N

              ——————— P
      O ———

    Q ———————
              ——— R

S ——————————— T

5 Use a ruler and a set square to draw a rectangle with a length of 8 cm and a width of 3 cm. Use the given line for one side.

Challenge

6 Draw a line parallel and a line perpendicular to the given lines through the points. To draw the lines without a set square, count the number of squares horizontally and vertically from one intersection to another on the grid.

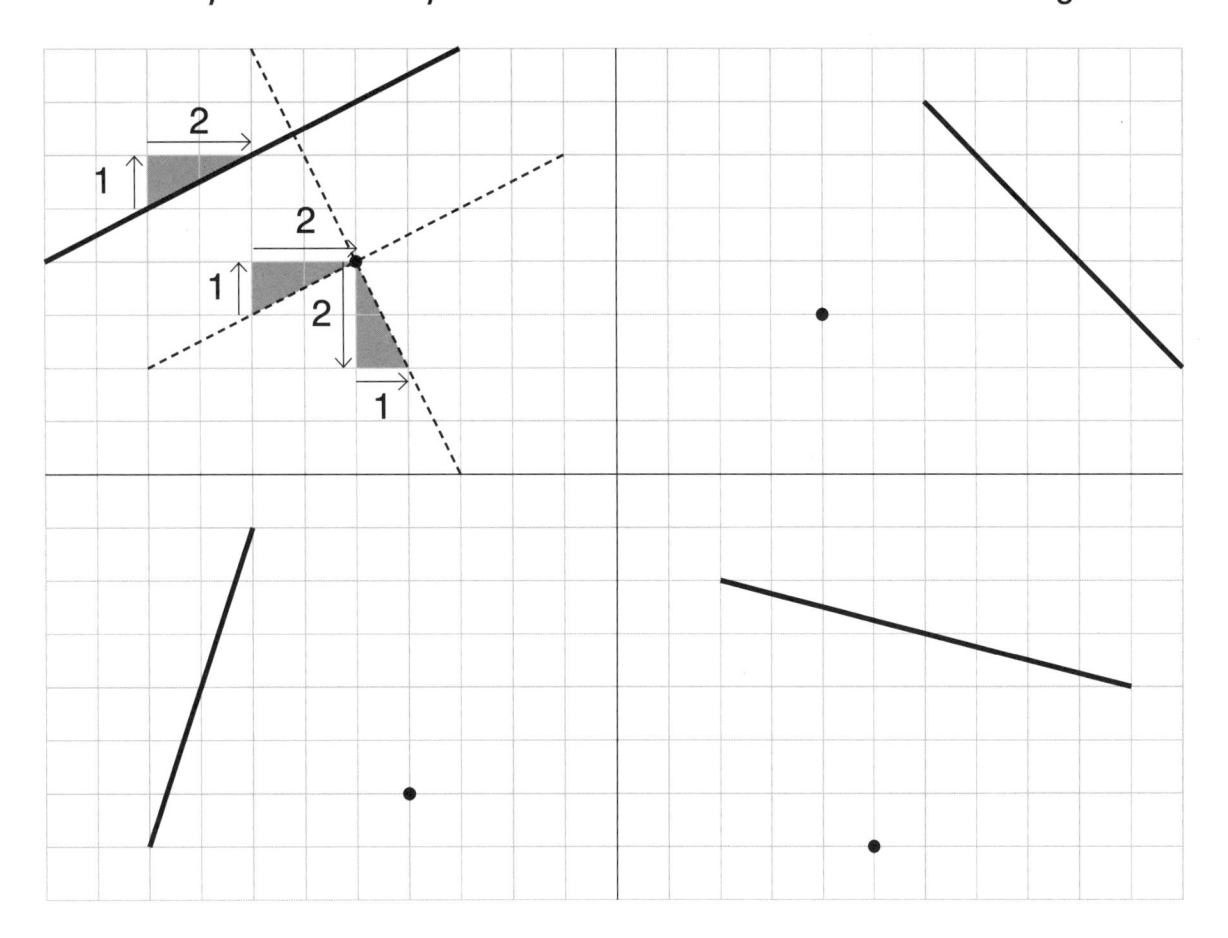

7 Draw a rectangle using the given line as one side. One side must pass through the given point.

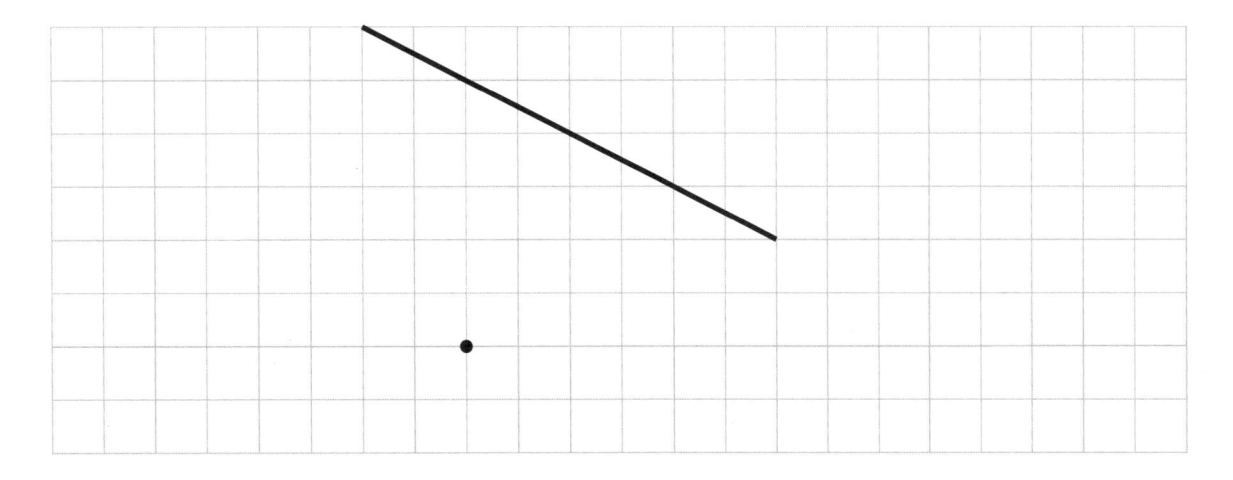

16-3 Drawing Perpendicular and Parallel Lines

Exercise 4

Basics

1 Write a check mark below each name for each quadrilateral. Use the definitions below.

	Trapezoid	Parallelogram	Rhombus	Rectangle	Square

Trapezoid   Quadrilateral with at least one pair of parallel sides.
Parallelogram   Trapezoid with two pairs of parallel sides.
Rhombus   Parallelogram with four equal sides.
Rectangle   Parallelogram with four right angles.
Square   Rhombus with four right angles.

Practice

2 Complete the table. Use a set square if needed to check right angles and parallel sides, and a ruler or compass to compare lengths.

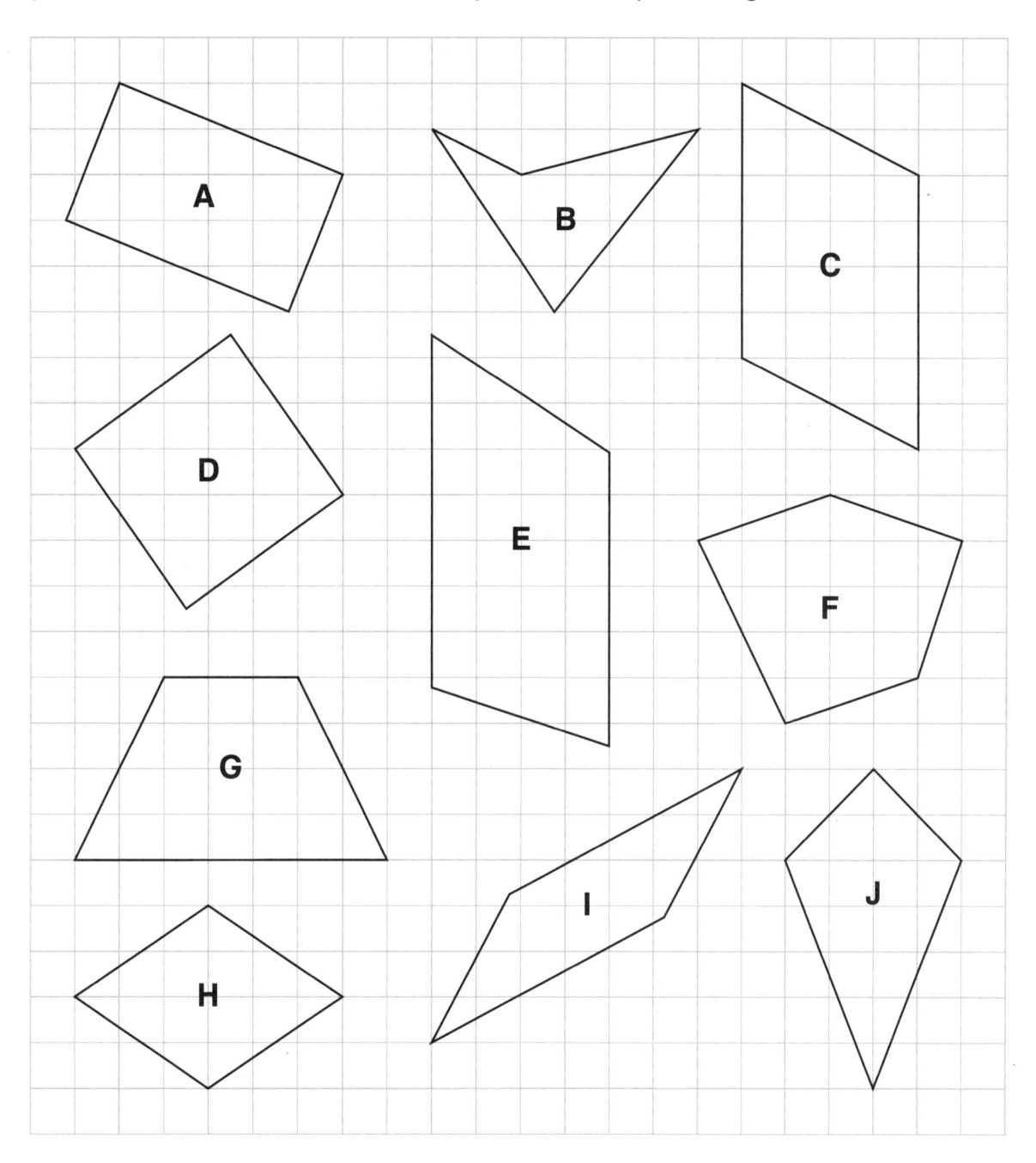

Quadrilateral	Trapezoid	Parallelogram	Rhombus	Rectangle	Square
A,	A,	A,		A,	

16-4 Quadrilaterals

3. Draw two different trapezoids that are not parallelograms, and two different parallelograms that are not rectangles.

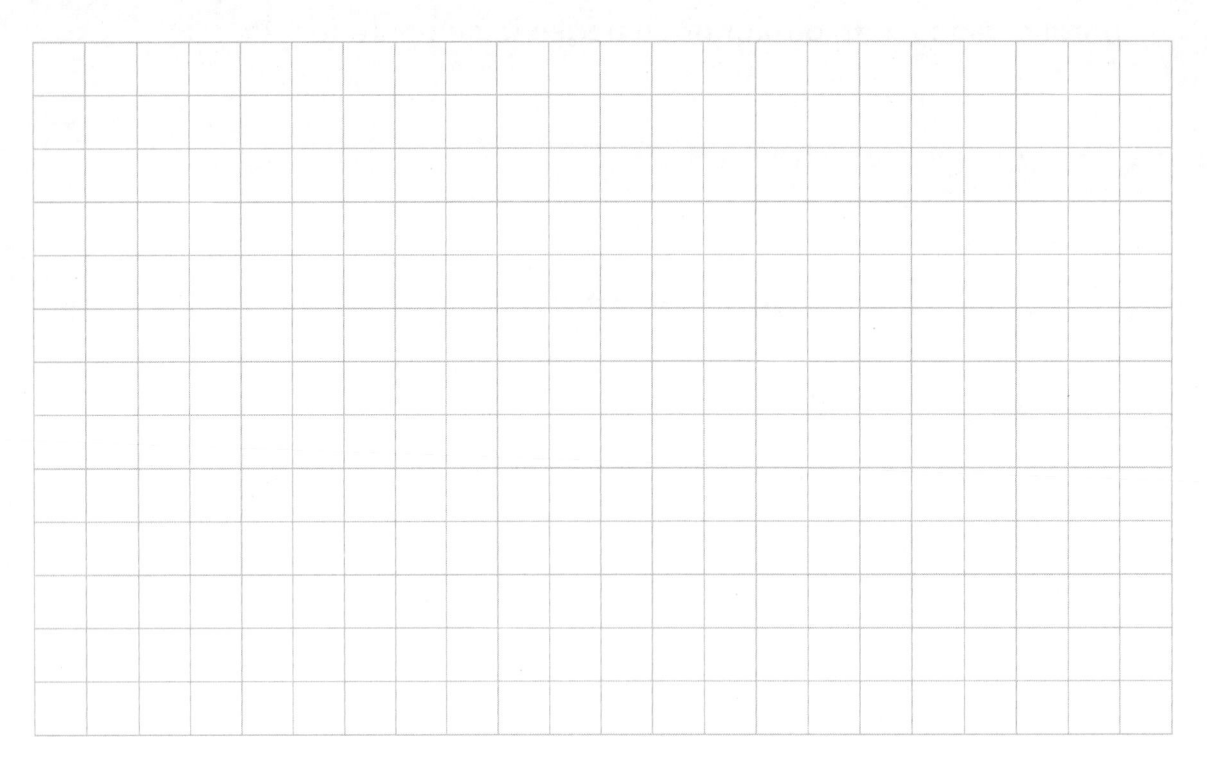

4. Name all the trapezoids in the figure below. Which of them are parallelograms? Which one is a rhombus?

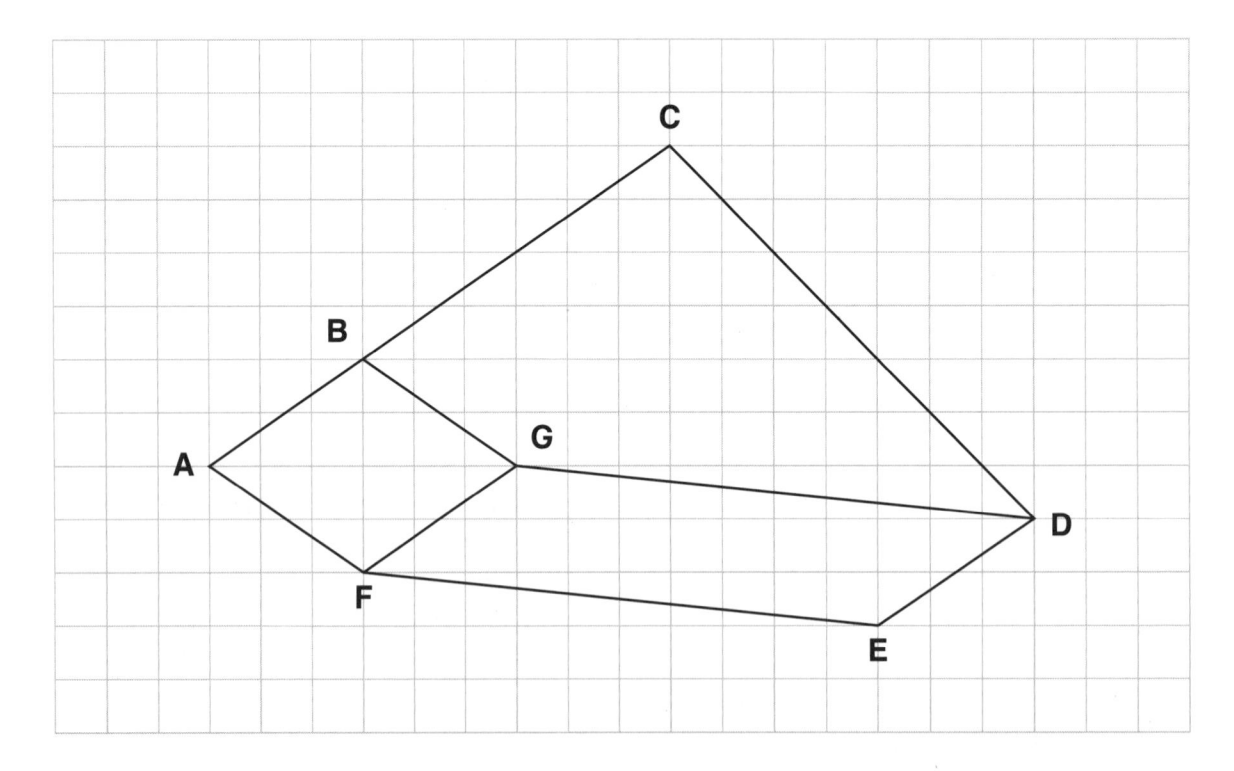

168   16-4 Quadrilaterals

Challenge

5 AC and BD are diagonals of Quadrilateral ABCD. They intersect at Point E. Use a compass or ruler to compare the lengths of AE to EC and DE to EB. What do you notice? This is a property of all parallelograms.

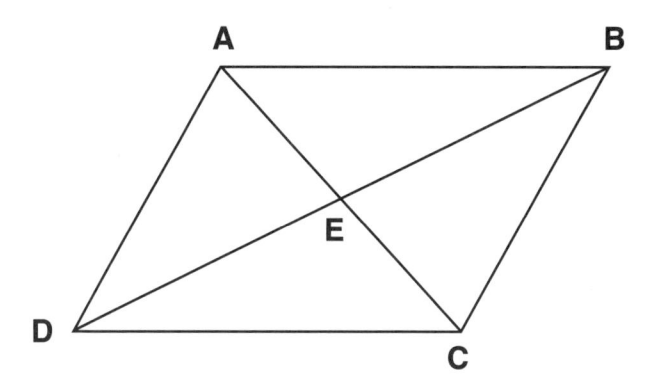

6 These two perpendicular lines intersect at their midpoint. Use them to draw a parallelogram. What kind of parallelogram is it?

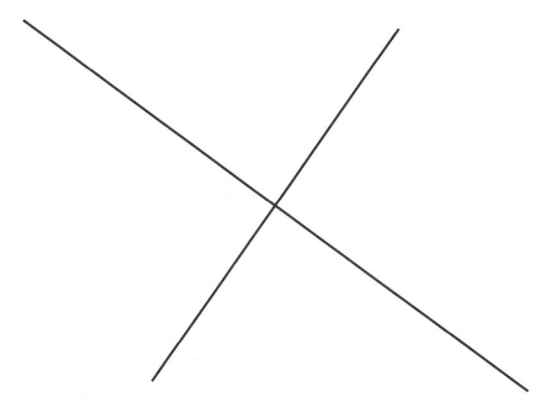

7 Use a ruler to draw a parallelogram by drawing the diagonals first.

Exercise 5

Basics

1 On which of these identical figures is the dotted line a line of symmetry? You can trace, cut out, and fold one to check.

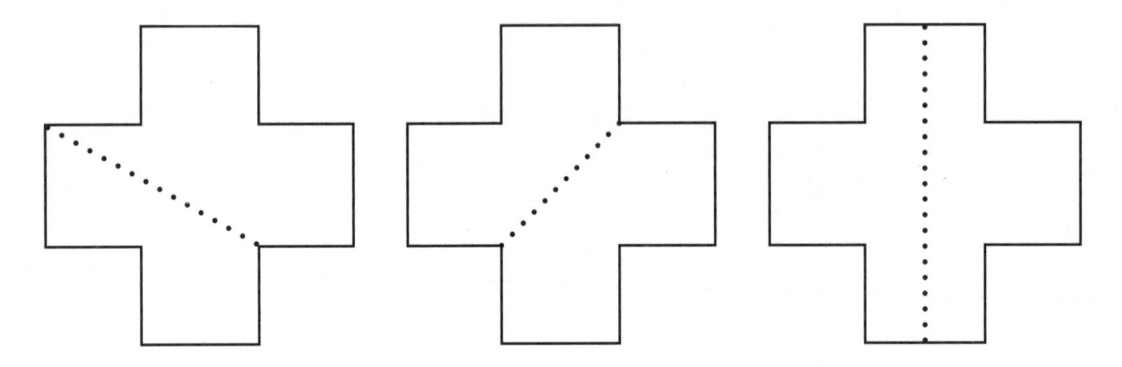

2 Check (✓) the box if the dotted line is a line of symmetry.

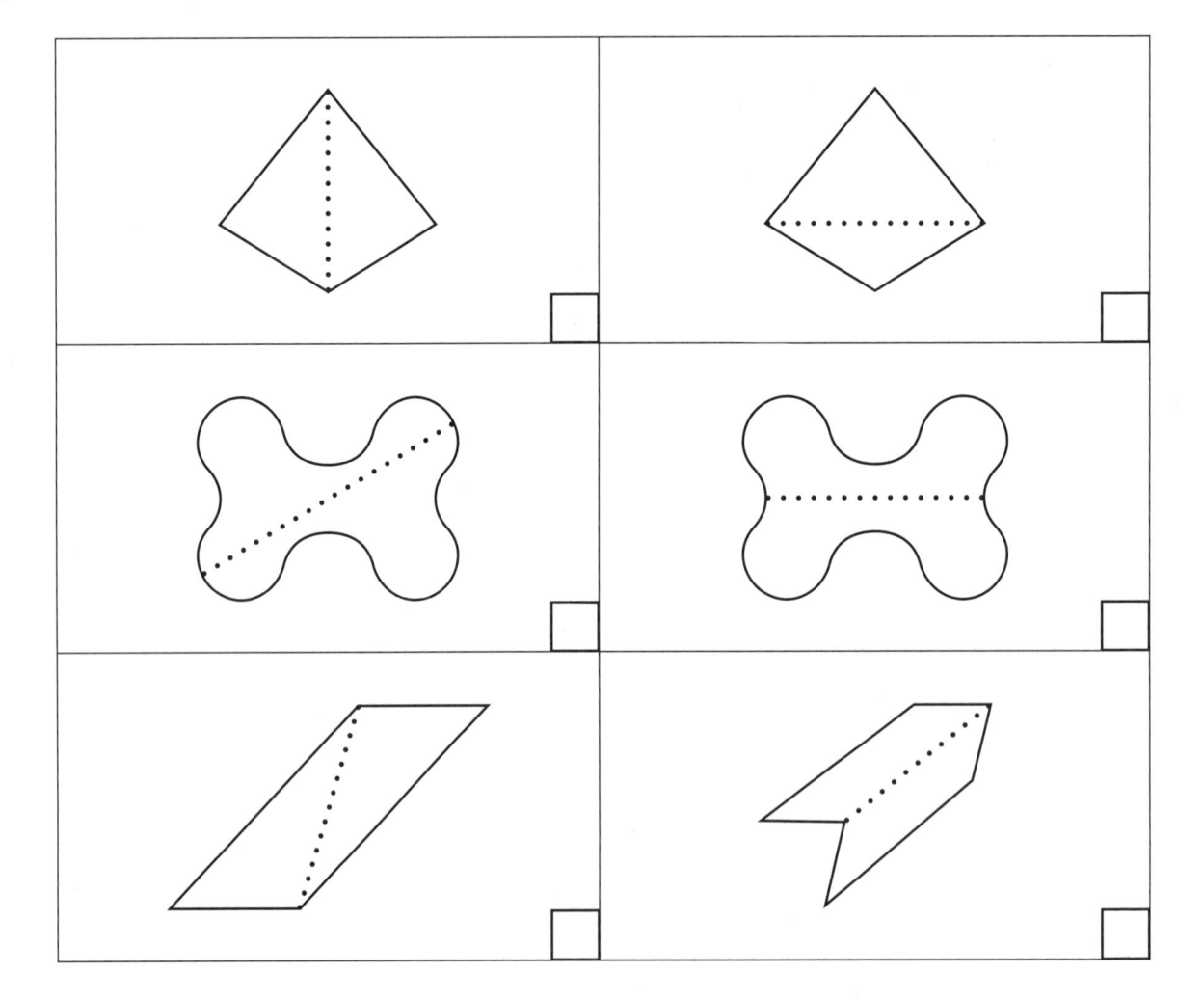

3️⃣ How many lines of symmetry do the equilateral triangle and the square each have? Draw them.

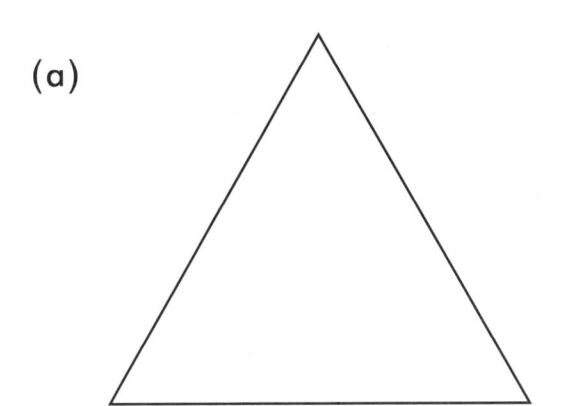

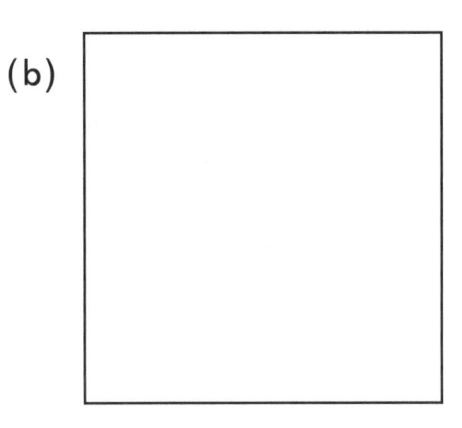

(a)

(b)

Practice

4️⃣ All the sides of the pentagon and the hexagon below are equal. How many lines of symmetry does each shape have? Draw them.

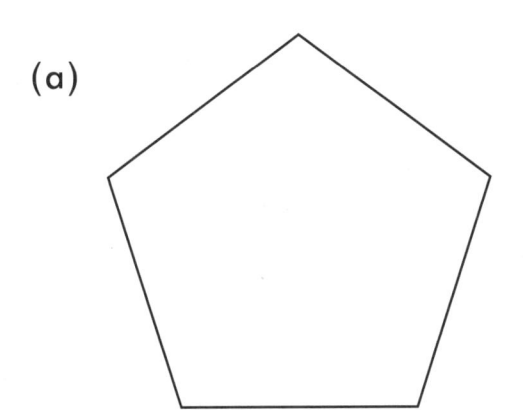

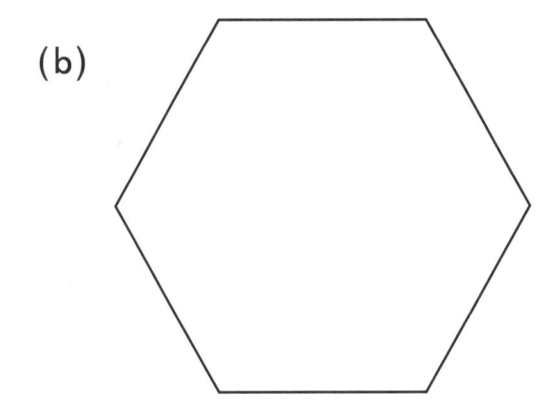

(a)

(b)

5️⃣ Find all the lines of symmetry in the following figures.

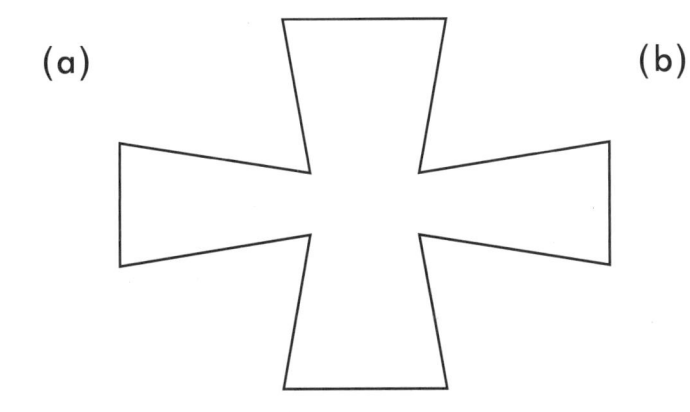

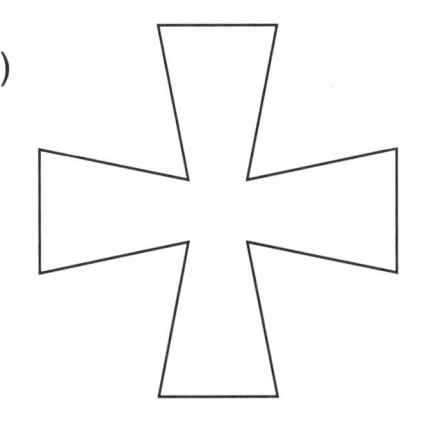

(a)

(b)

6 For each figure, if it is symmetrical, draw a line of symmetry.

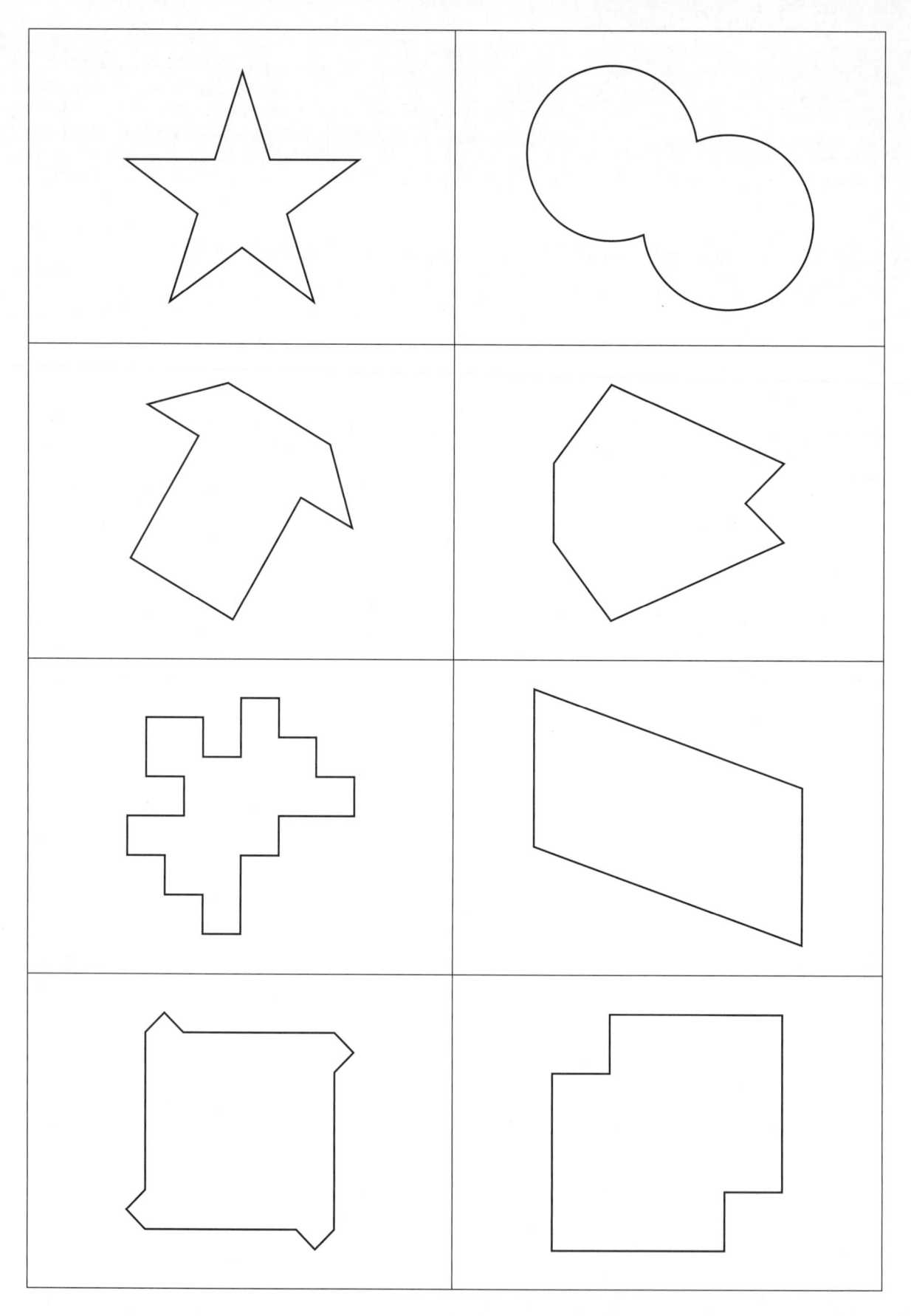

Challenge

7 The dotted line is a line of symmetry for the entire figure.

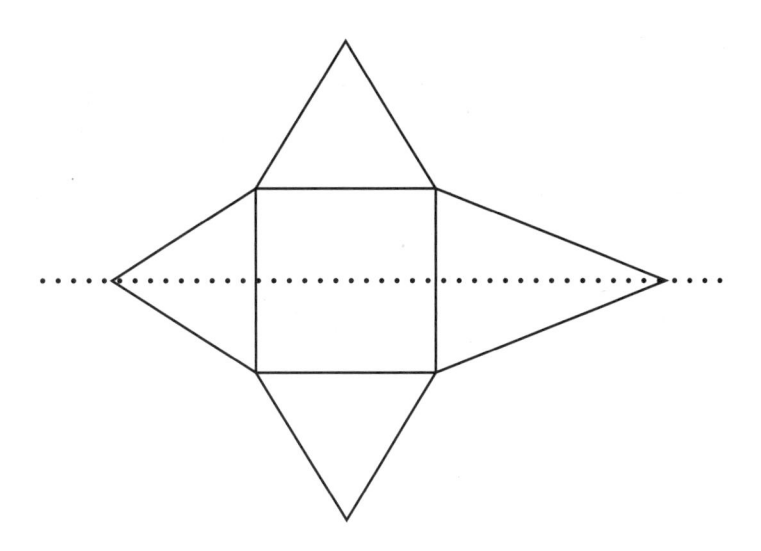

Draw a line of symmetry for the following figures.

(a)

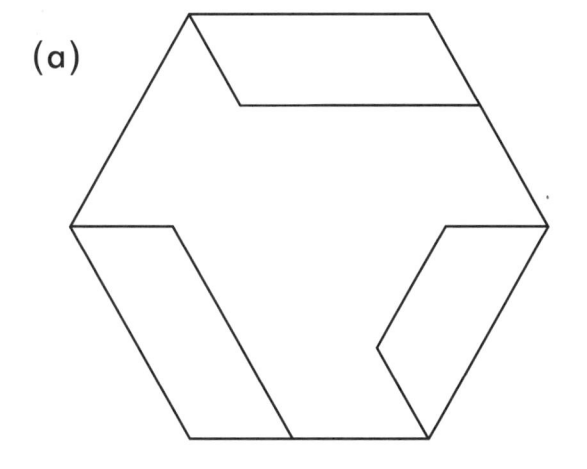

(b)

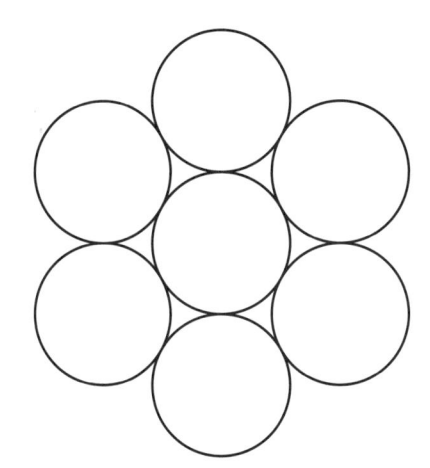

(c)

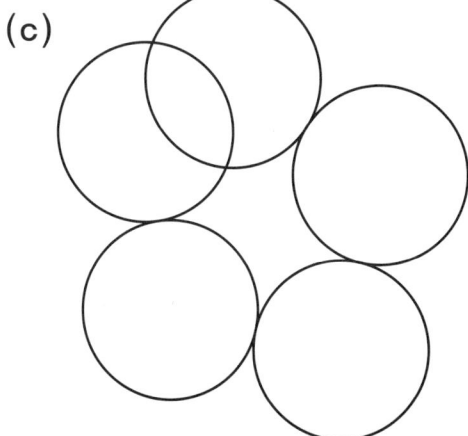

(d)

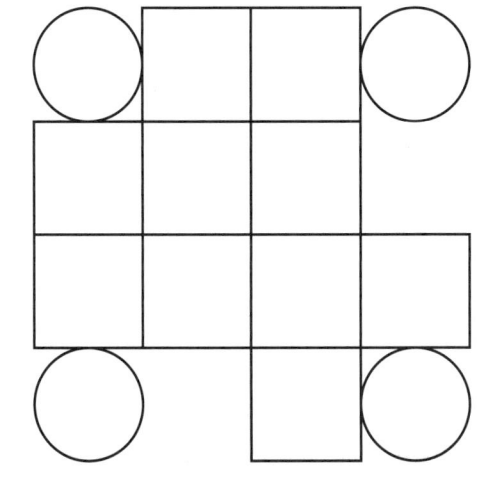

16-5 Lines of Symmetry

Exercise 6

Basics

1 Use the dots and a ruler to complete the symmetrical figures. Each dot is the same perpendicular distance from the line of symmetry as a vertex on the other side of the line of symmetry.

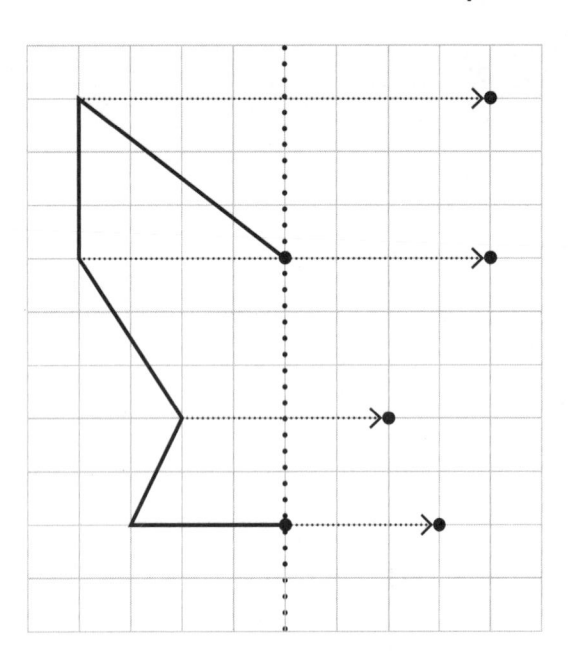

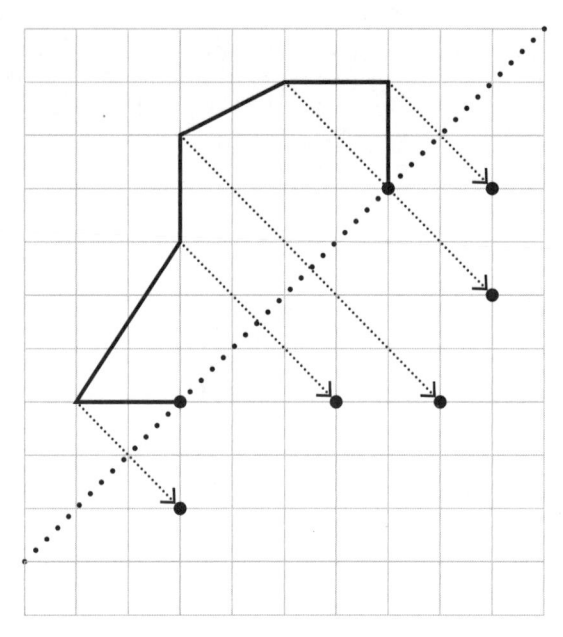

2 Complete the symmetrical figures by shading three more squares on each grid.

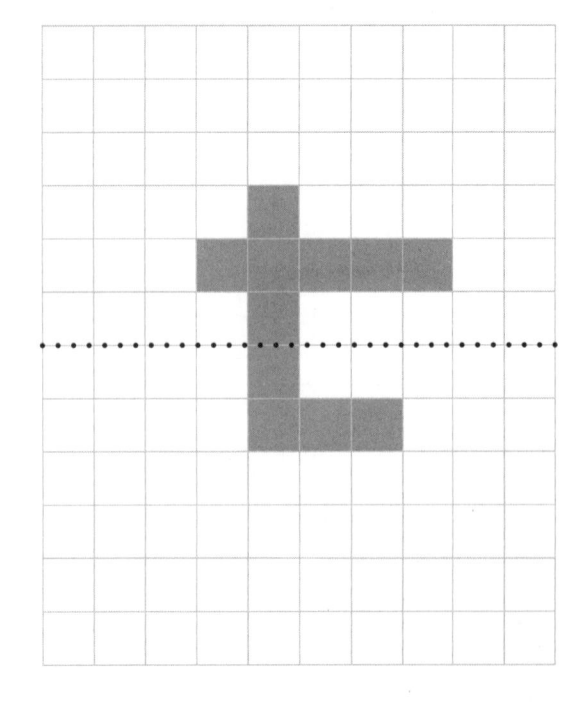

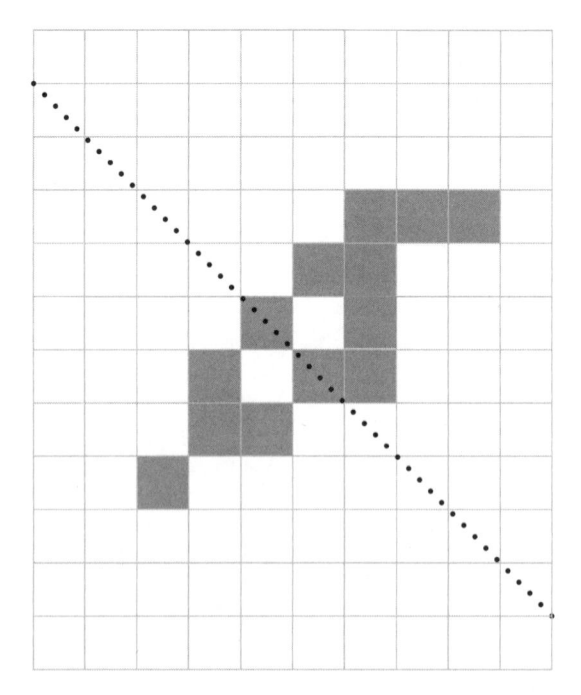

16-6 Symmetrical Figures and Patterns

Practice

3 Complete each figure so that the dotted line is a line of symmetry.

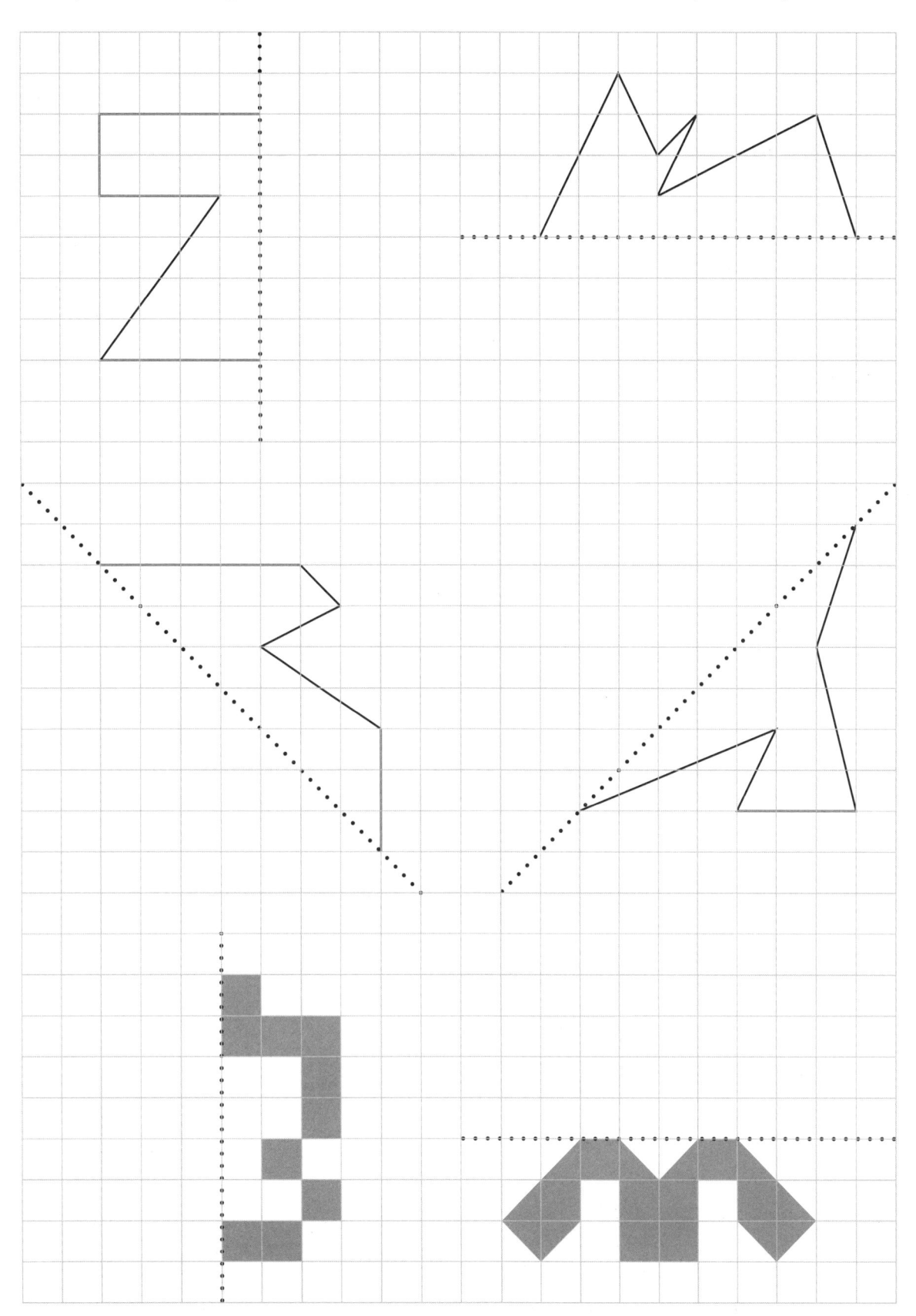

Challenge

4

176 16-6 Symmetrical Figures and Patterns

Exercise 7

Check

1 In this figure, "S" is written inside a square.

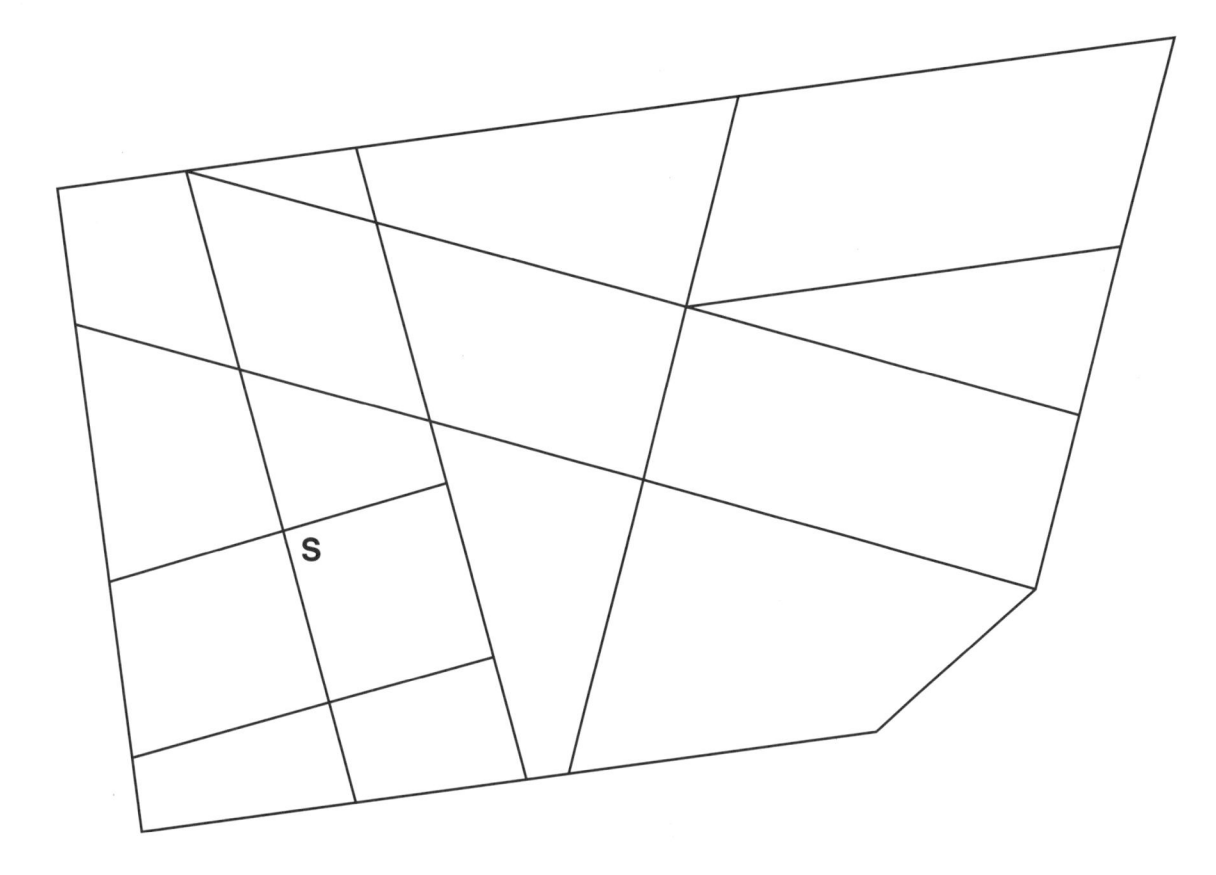

(a) Write "Rh" inside the rhombuses.

(b) Write "Re" inside the rectangles.

(c) Write "P" inside the parallelograms.

(d) Write "T" inside the trapezoids.

(e) Write "Q" inside the quadrilaterals.

2 Use a ruler and a set square to draw one line perpendicular and one line parallel to the given line through each of the given points.

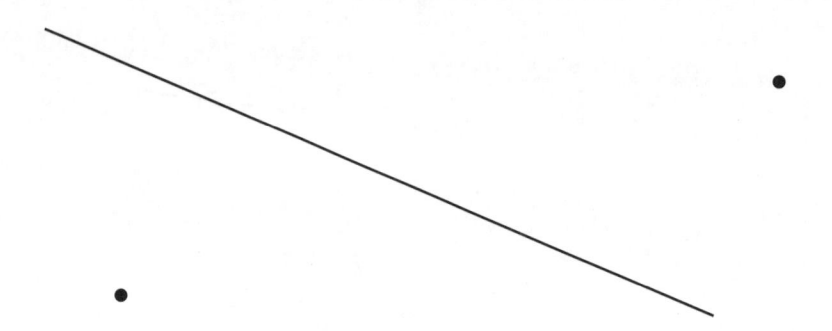

3 Use a ruler and set square to draw a parallelogram that is not a rectangle.

4 Use a ruler and set square to draw a trapezoid that is not a parallelogram.

5 (a) What type of quadrilateral is the symmetrical figure below?

(b) Draw a line of symmetry and write the measure of ∠a.

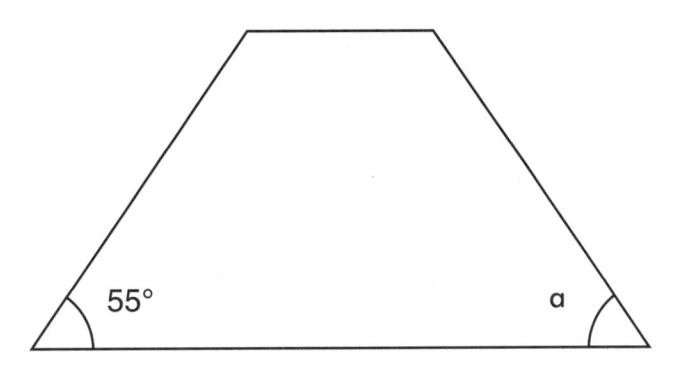

55°   a

6 Draw the lines of symmetry on the equilateral triangle below. Write the measure of the other two angles. (Use symmetry.)

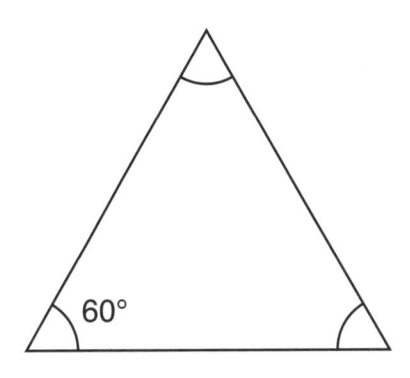

60°

7 Draw a line of symmetry on the isosceles triangle below. Write the measure of ∠a.

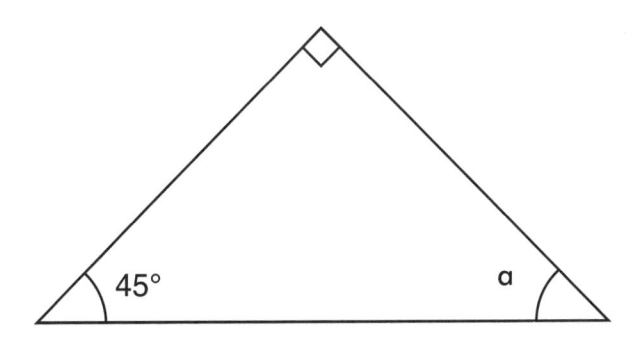

45°   a

8 Complete each figure so that the dotted line is a line of symmetry.

(a)

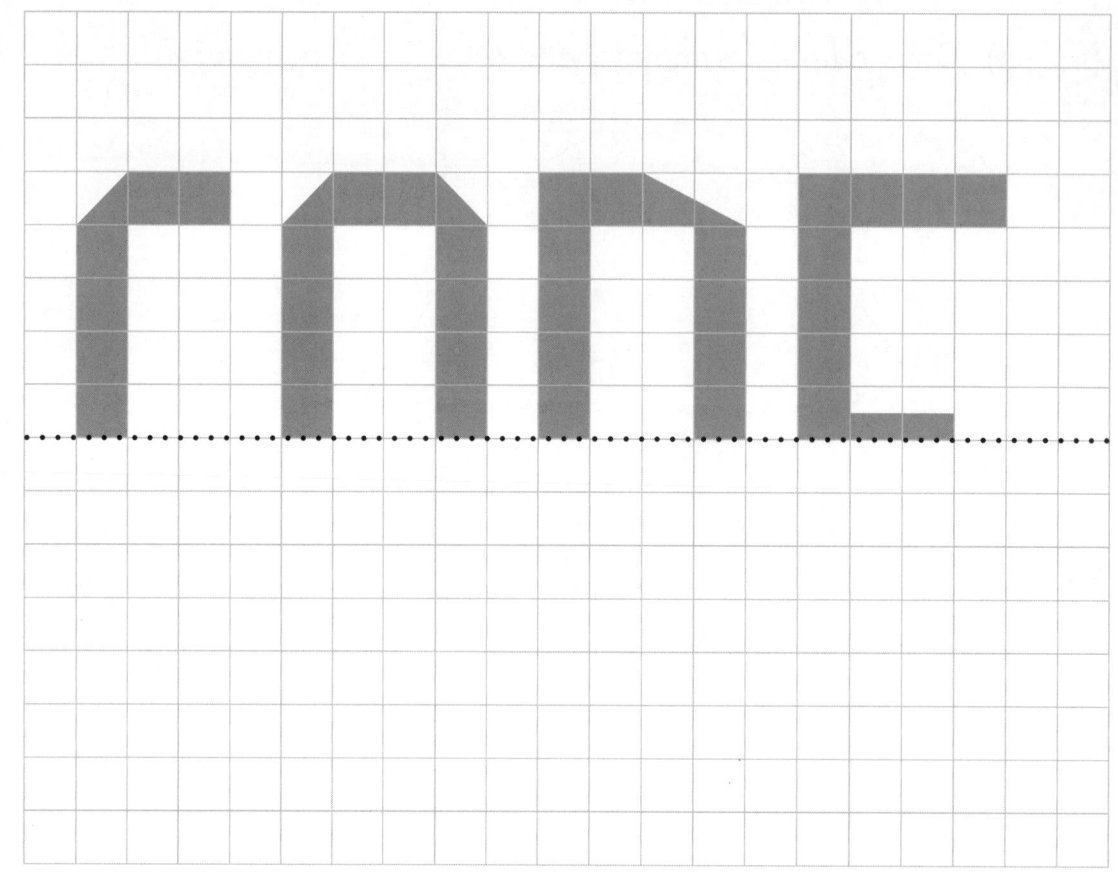

(b)

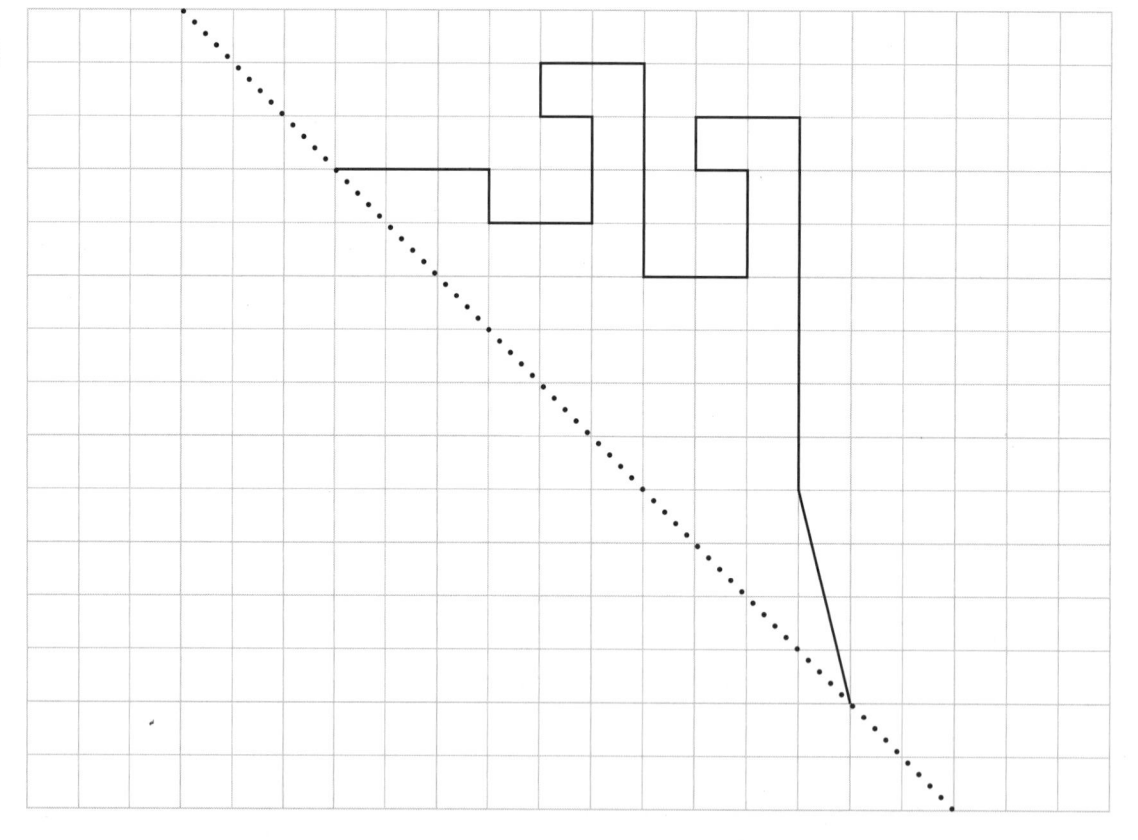

Challenge

9 What word will be formed if the figure below is completed so that the dotted line is a line of symmetry?

10 Shade one more square in the entire figure to make it symmetrical.

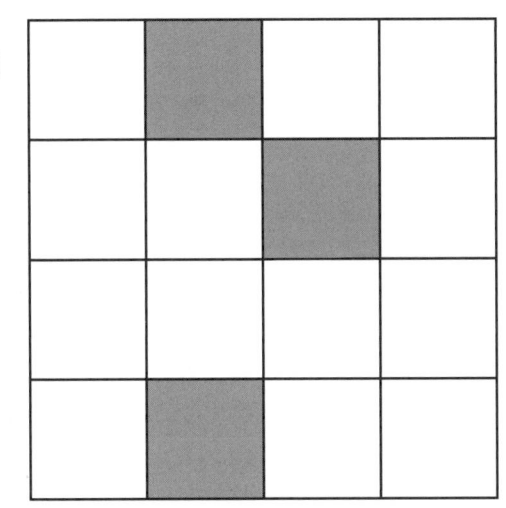

(a)

(b)

11 Complete the figure so that it has two lines of symmetry.

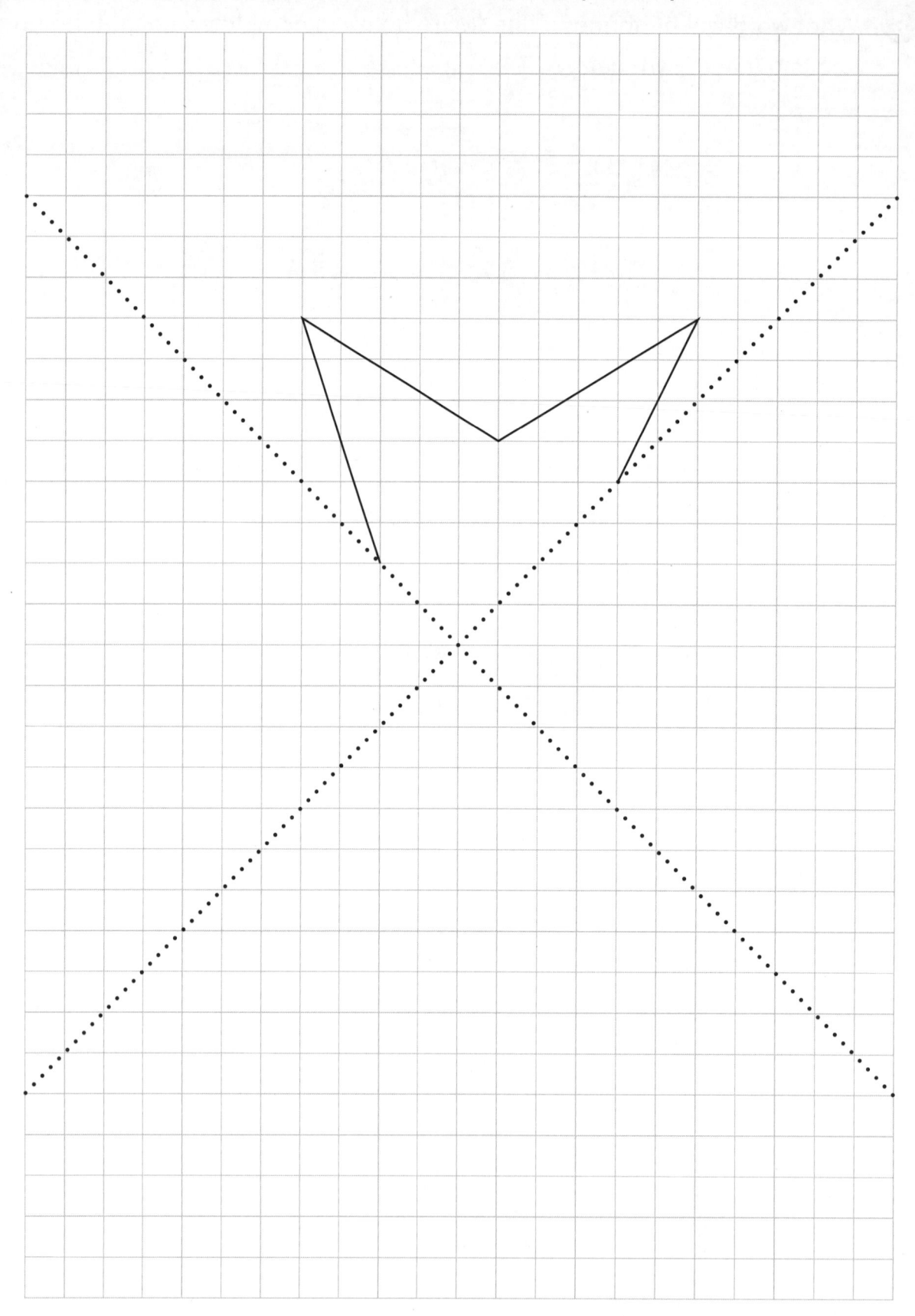

182   16-7 Practice

Chapter 17 Properties of Cuboids

Exercise 1

Basics

1 These are illustrations of the same cuboid.

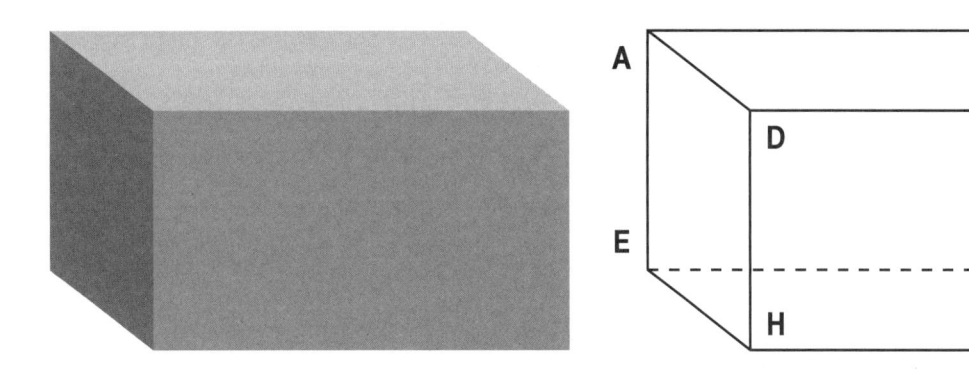

(a) A cuboid has _____ faces, _____ vertices, and _____ edges.

(b) Face ABCD is the same size and shape as Face _____.

(c) Name two other pairs of faces that have the same size and shape.

(d) On Face ABCD, AD is the same length as _____.

(e) List the other edges that have the same length as AD.

(f) List all the edges with the same length as AB.

(g) List all the edges with the same length as AE.

Practice

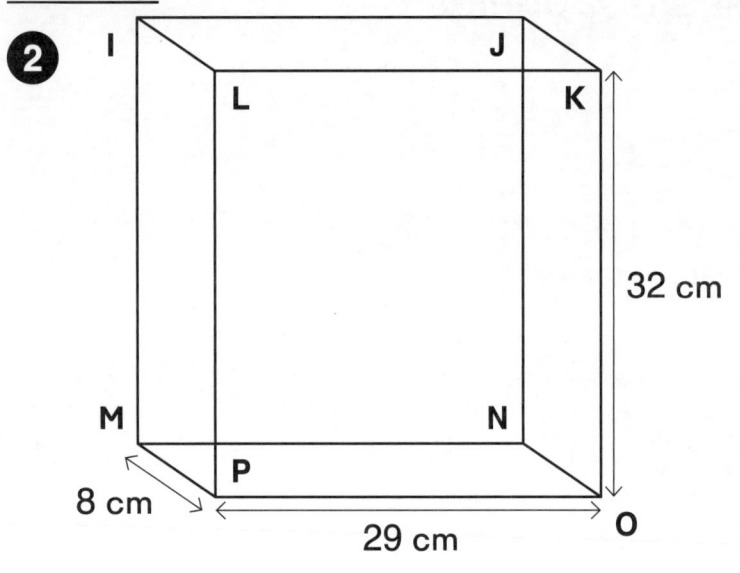

2

(a) Which edges have the same length as NO?

(b) Which edges have the same length as PO?

(c) Which edges have the same length as JN?

(d) Find the area of each of the faces.

3 For this cube, the length of Edge RS is 9 cm.

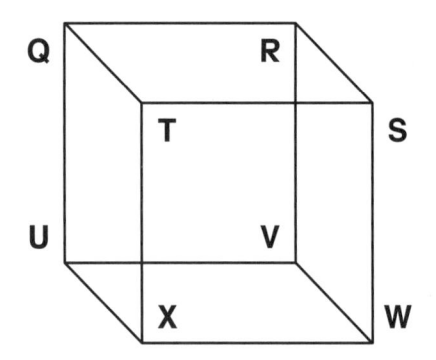

(a) What is the length of Edge UV?

(b) What is the area of Face QTXU?

4 For this cuboid the area of Face WZDA is 48 cm².

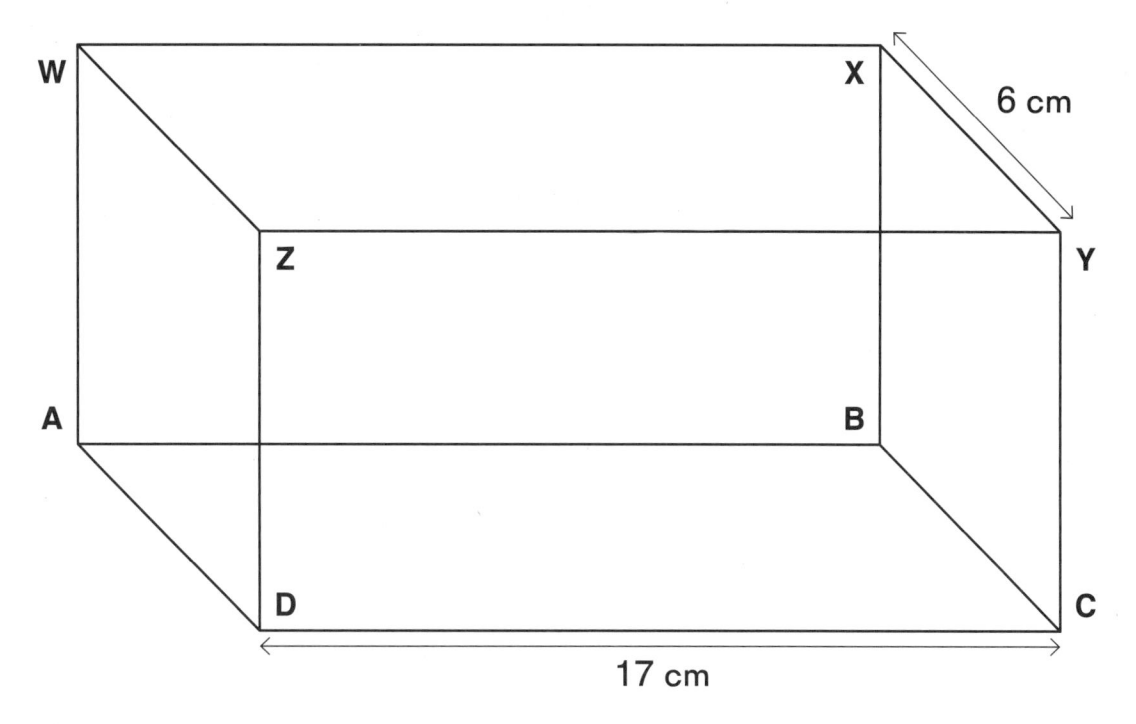

(a) What is the area of Face WXBA?

(b) What is the perimeter of Face WZDA?

5 For this cuboid, the perimeter of Face EFGH is 98 cm.

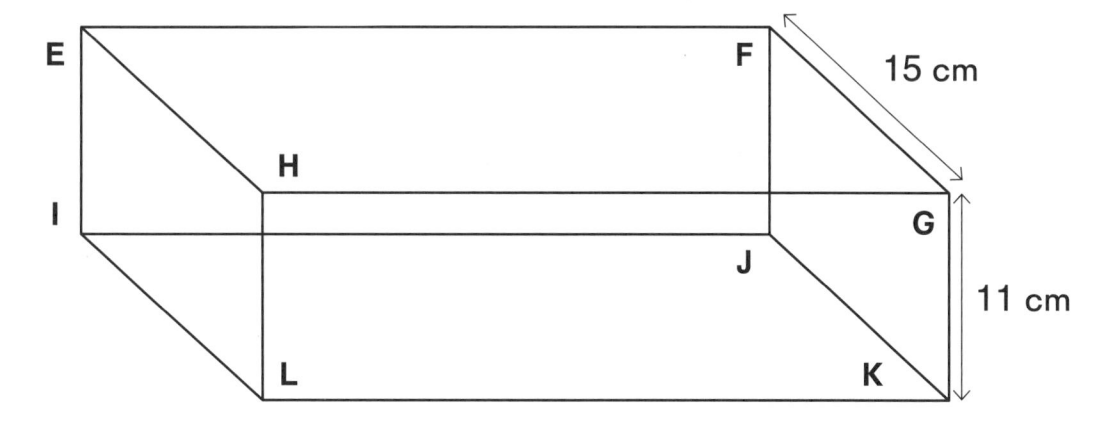

(a) What is the area of Face HGKL?

(b) What is the perimeter of Face IJKL?

17-1 Cuboids   185

Exercise 2

Basics

1 A and C are nets of this cuboid, but B and D are not. Explain why B and D cannot be nets of this cuboid.

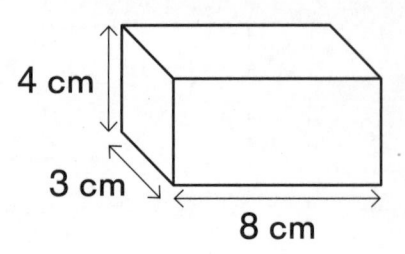

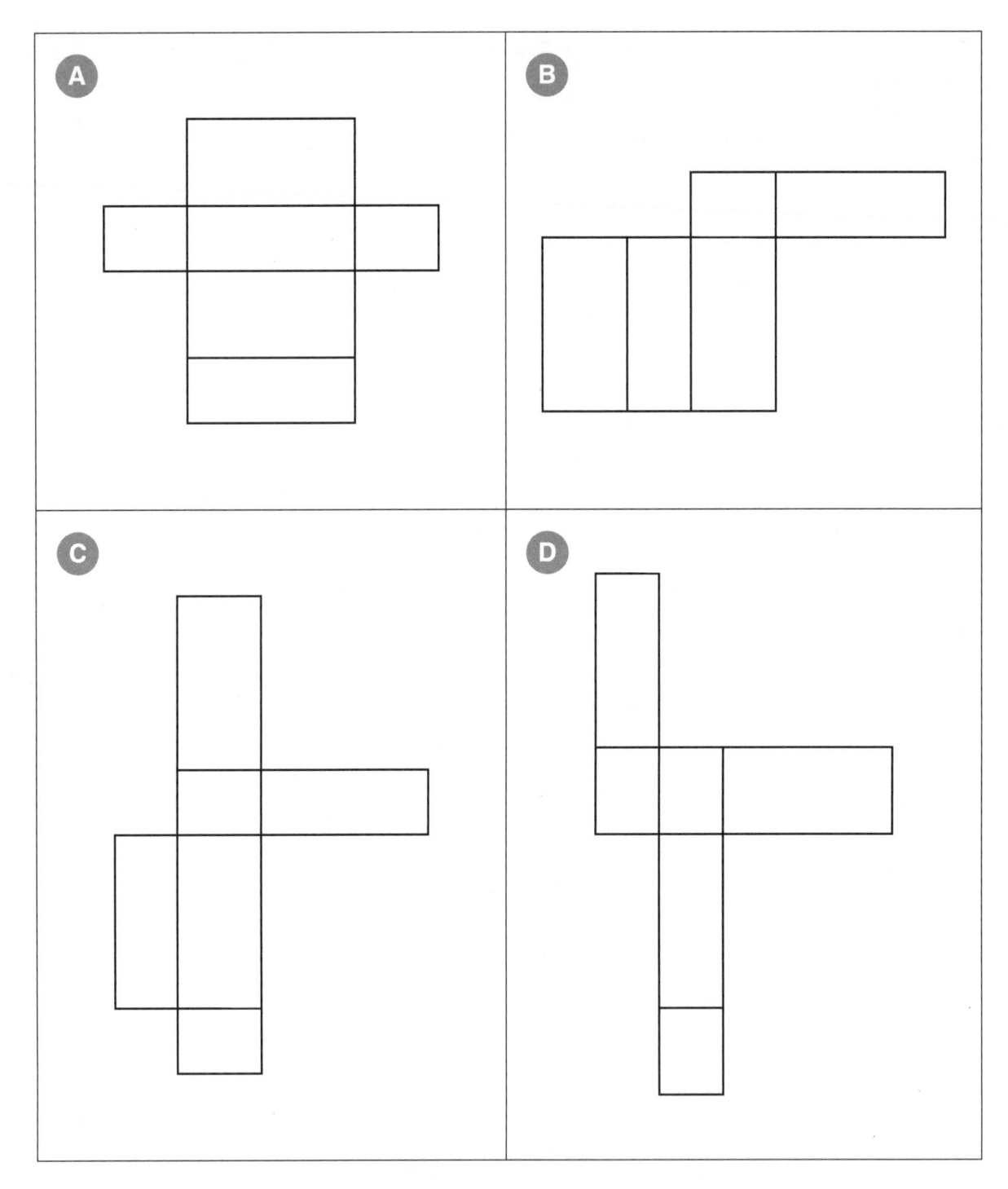

4 cm
3 cm
8 cm

A

B

C

D

186   17-2 Nets of Cuboids

Practice

2 Label the rest of the vertices of the two nets of the cuboid to match the vertices of the cuboid. Some letters will be used more than once. (You can copy the nets onto graph paper and fold them if needed.)

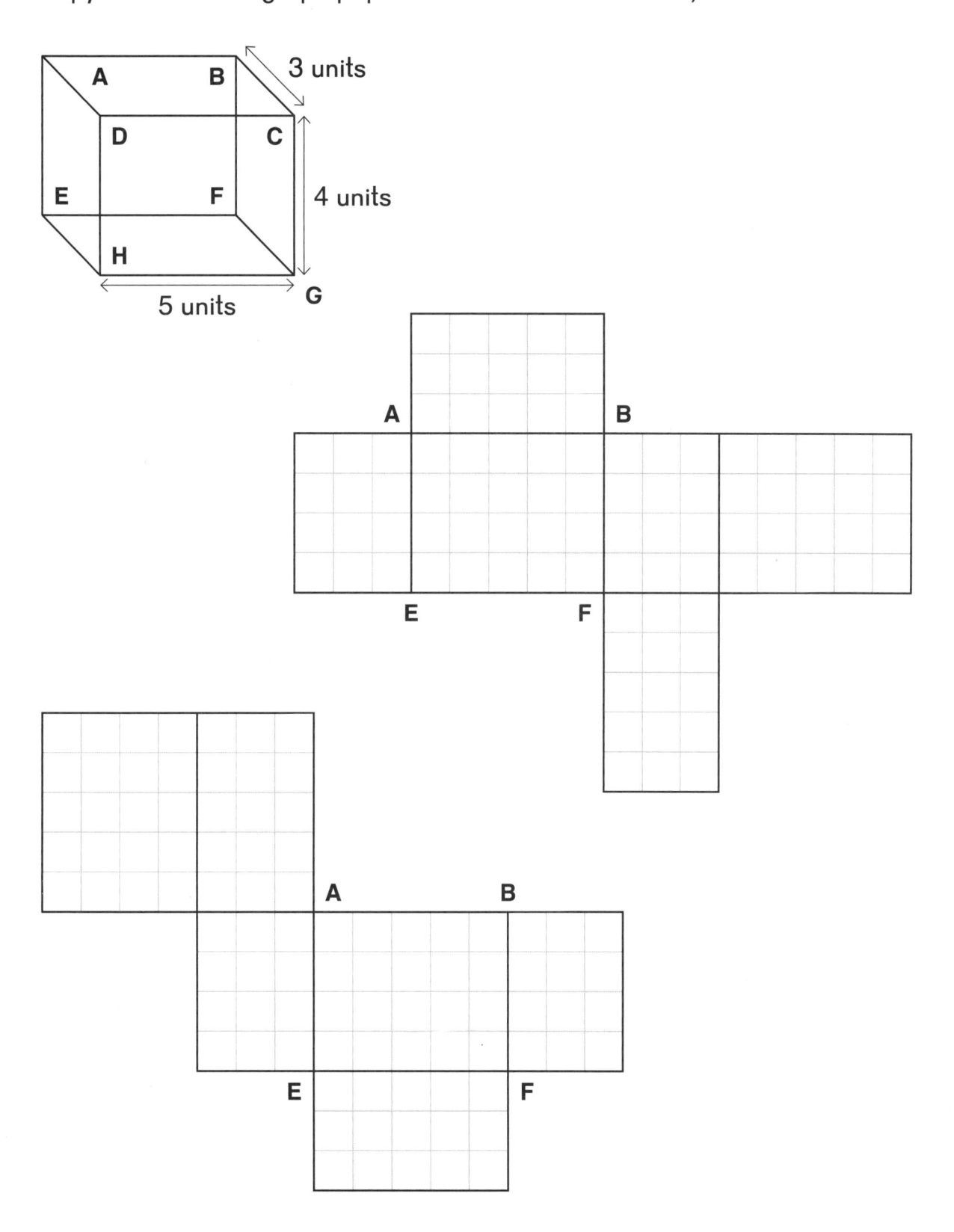

17-2 Nets of Cuboids

3 Draw more faces on each of the following nets so that they can be folded to form the given cube. Each net should be different.

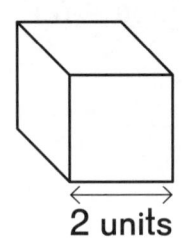

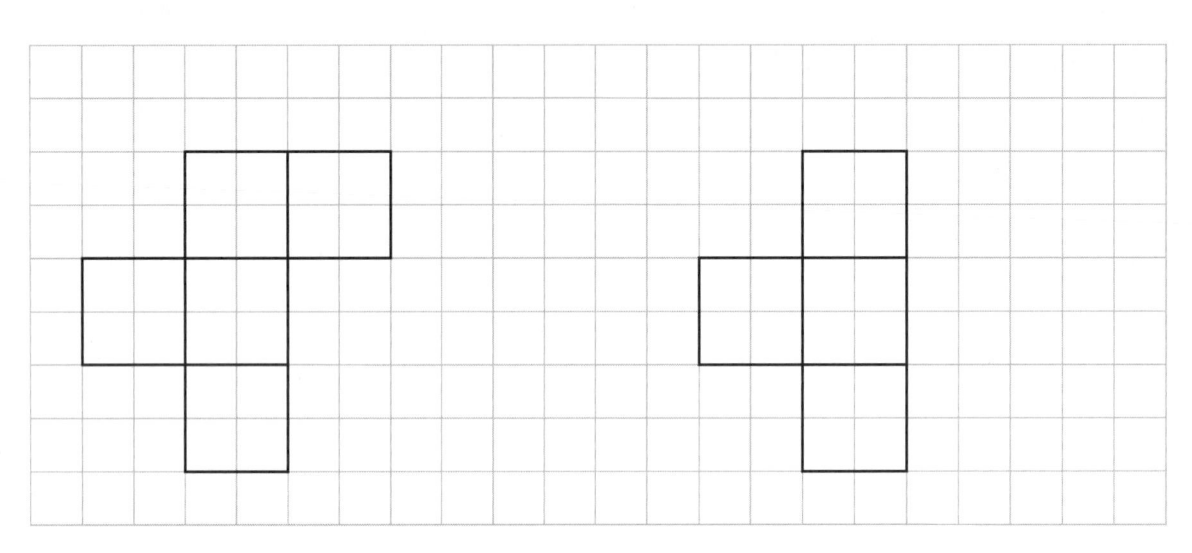

4 Draw more faces on each of the following nets so that they can be folded to form the given cuboid.

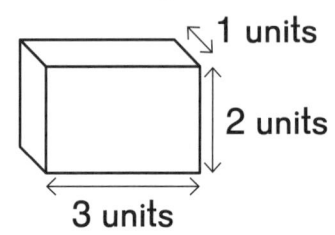

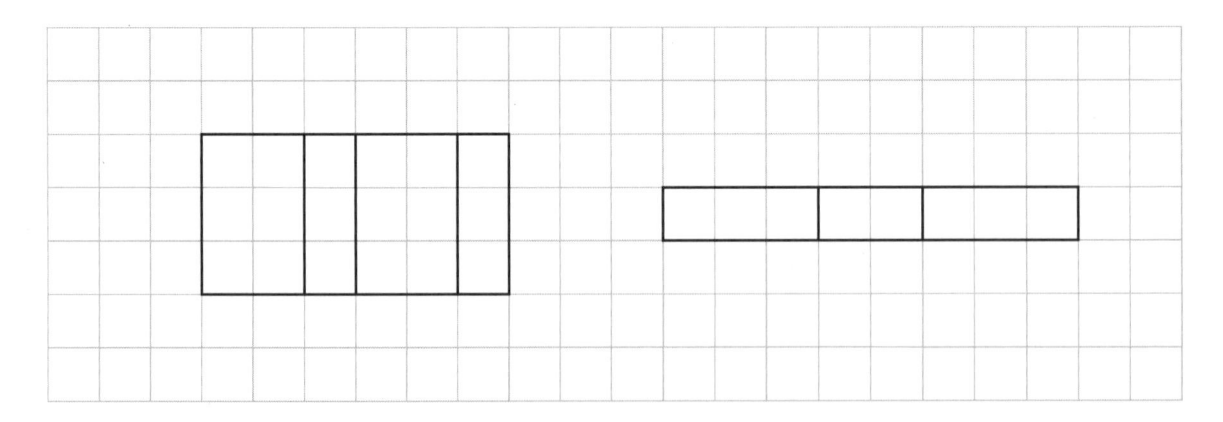

Challenge

5 The faces of this cube are numbered 1 through 6 so that the sum of the numbers on opposite sides is always 7.

Number the rest of the faces of each of these nets so that when they are folded into a cube, the sum of the numbers on opposite sides will be 7.

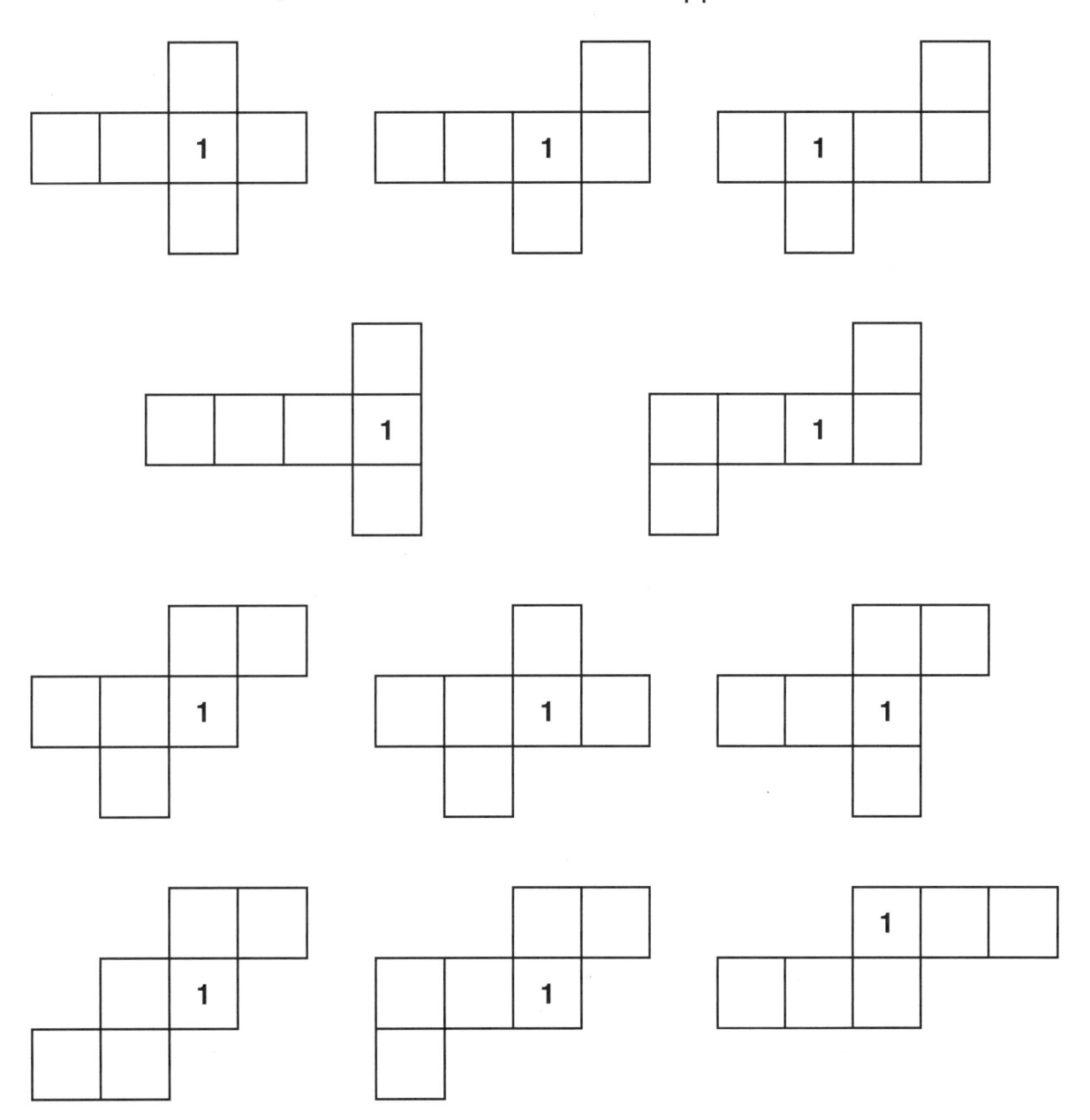

17-2 Nets of Cuboids

Exercise 3

Basics

1 These are illustrations of the same cuboid.

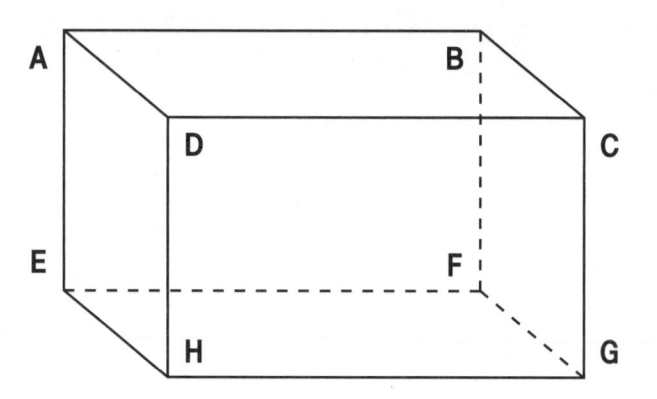

(a) Face ABCD is parallel to Face _____.

(b) Name two other pairs of faces that are parallel to each other.

(c) Face ABCD and Face _____ are perpendicular to each other at Edge DC.

(d) Name three other faces that are perpendicular to Face ABCD.

(e) Name all the faces that are perpendicular to Face ADHE.

190   17-3 Faces and Edges of Cuboids

(f) On Face ABCD, Edge AD is parallel to Edge _____.

(g) On Face ADHE, Edge AD is parallel to Edge _____.

(h) Name another edge that is parallel to Edge AD.

(i) Name three edges that are parallel to Edge AE.

(j) Name two edges that are perpendicular to Edge AD.

(k) Name four edges perpendicular to Edge HG.

(l) Name three edges that are not perpendicular to Edge FG.

Practice

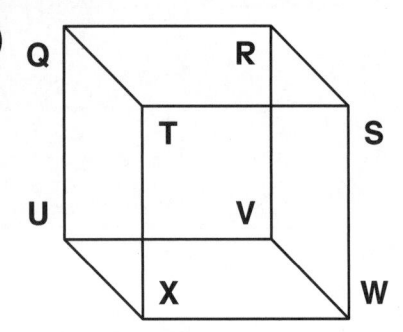

2

(a) Name two edges that are parallel to Edge XW.

(b) Name two edges that are perpendicular to Edge QR.

(c) Which face is parallel to Face RSWV?

(d) Which faces are perpendicular to Face RSWV?

3 Draw two more cuboids below. Each cuboid should be different.

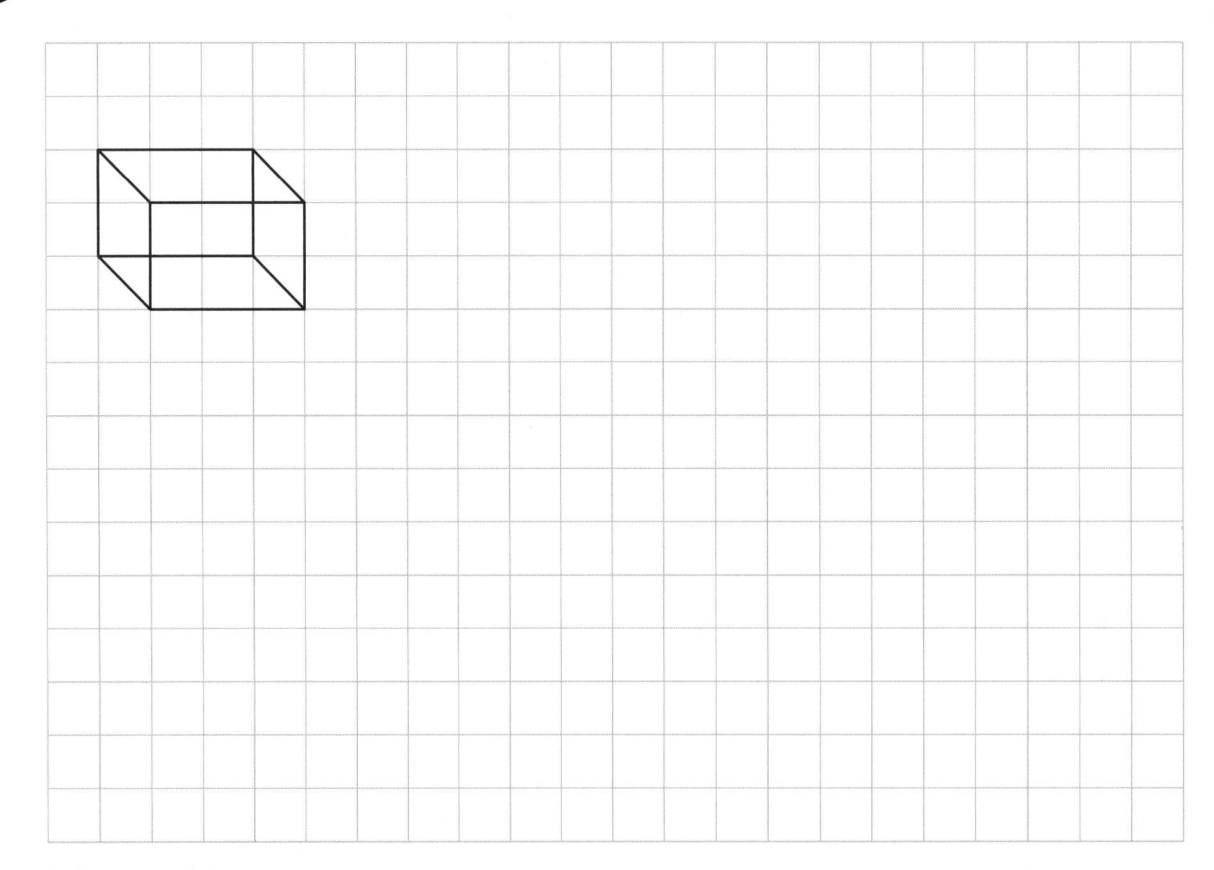

4 A cuboid is formed from this net.

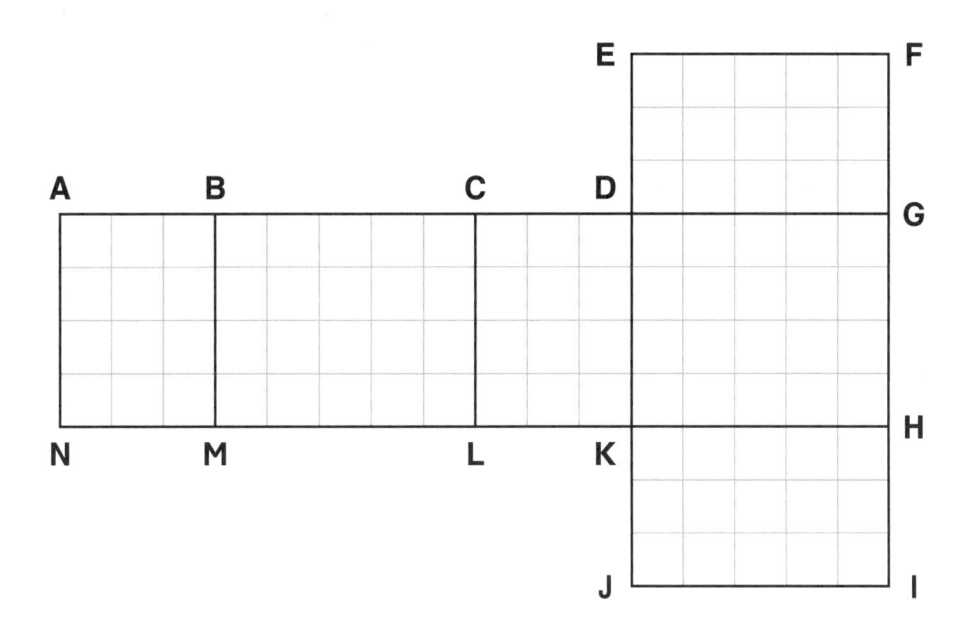

(a) List all pairs of faces that will be parallel to each other.

(b) Name the faces that will be perpendicular to Face BCLM.

Challenge

5 For each of these nets, shade the four faces that will be perpendicular to the shaded face when the net is folded.

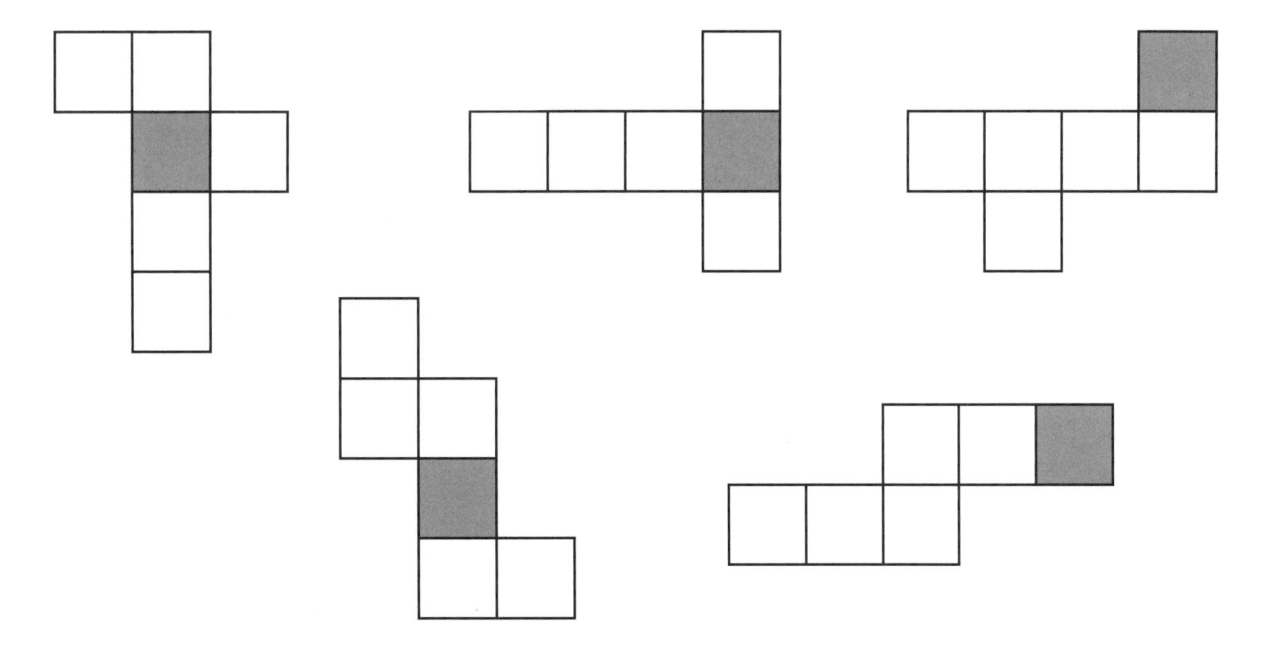

Exercise 4

Check

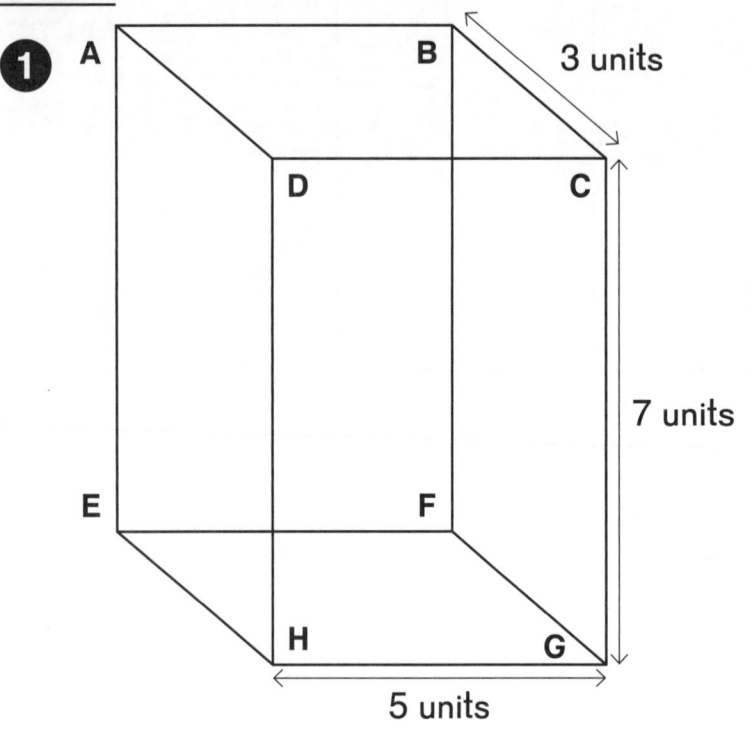

1. [Cuboid ABCD-EFGH with AB = 5 units (labeled as 5 units along HG), BC = 3 units, and height 7 units]

(a) Find the area of the face that is parallel to Face ABCD.

(b) Find the area of two faces that are perpendicular to Face EFGH and have D as one vertex.

(c) If another identical cuboid was glued to Face DCGH to make a larger cuboid, what would be the area of the top face of the new cuboid?

(d) On the next page, two different nets of the cuboid have been started. Complete each net and find the perimeter of each.

17-4 Practice

2. On this cuboid, Face TWSP is a square and has a perimeter of 32 cm. PQ is 12 cm long.

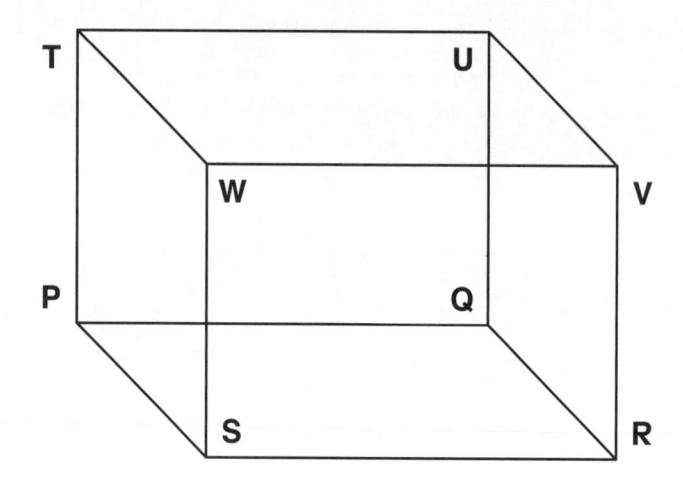

(a) Which other face is also a square?

(b) What is the length of Edge VR?

(c) What is the area of each face that is perpendicular to Face TWSP?

3. This is a net of a cube.

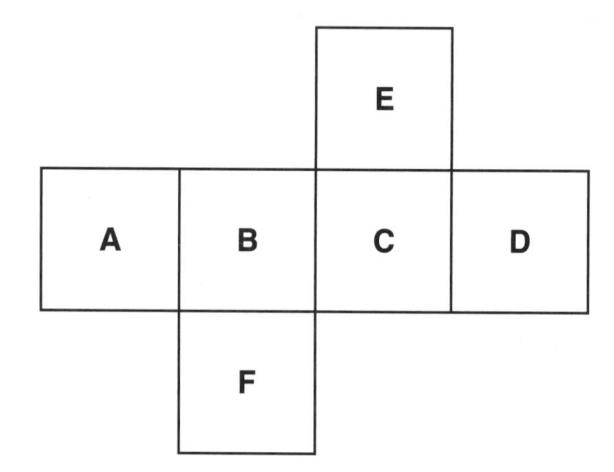

(a) When folded, which faces will be perpendicular to Face A?

(b) Which face will be parallel to Face B?

Challenge

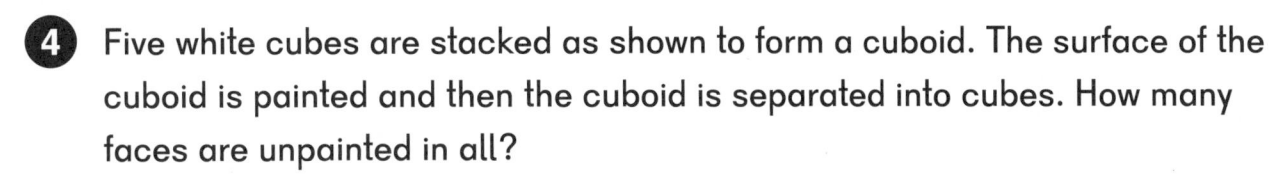

4 Five white cubes are stacked as shown to form a cuboid. The surface of the cuboid is painted and then the cuboid is separated into cubes. How many faces are unpainted in all?

5 64 white cubes are stacked as shown to form a large cube. The surface of the cuboid is painted and then the cuboid is separated into cubes.

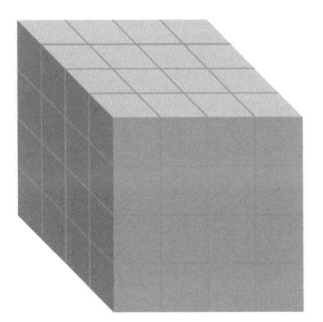

(a) How many cubes have exactly 3 faces painted?

(b) How many cubes have only 2 faces painted?

Exercise 5

Check

1. | 36.6 | 6.53 | 56.36 | 345.05 | 3.65 |

   (a) What does the digit 3 stand for in each number?

   (b) Arrange the numbers in order from least to greatest.

   (c) Write the greatest number in expanded notation.

   (d) Multiply the number with the digit 5 in the first decimal place by 6.

   (e) Divide the difference between the least number and the greatest number by 5.

   (f) Find the sum of the 3 least numbers.

2

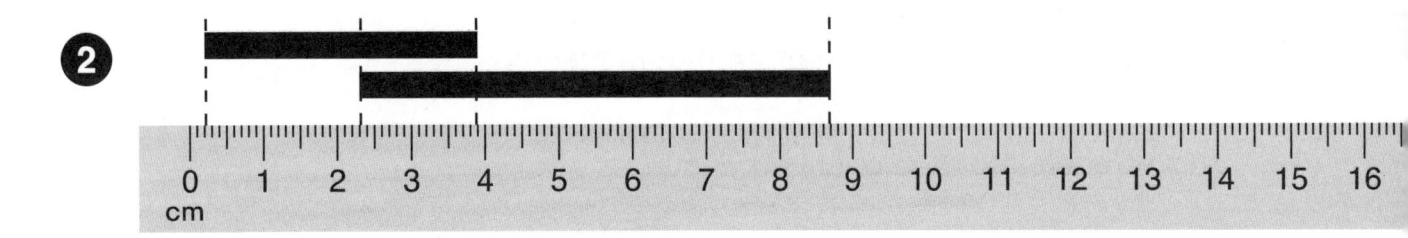

(a) Express the difference in length between the two lines as a decimal.

(b) Express the difference in length between the two lines in compound units of centimeters and millimeters.

(c) Express the sum of the lengths of the two lines as a mixed number in simplest form.

3 What is the difference in perimeter of these two figures?

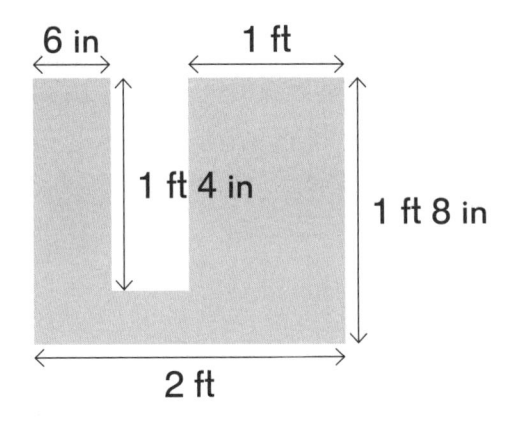

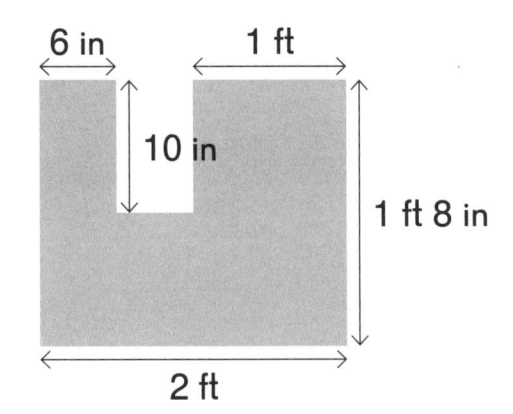

4 Martin has a rectangular garden that measures 6 ft by 12 ft.

(a) What is the area of his garden in square feet?

(b) What is the area of his garden in square yards?

(c) There is a 3-foot wide path all the way around Martin's garden. What is the total area of both the garden and the path in square feet?

5 (a) The table below shows the amount of water Martin used each month to water his garden from April to September. Complete the line graph to show this data.

Month	Apr	May	Jun	Jul	Aug	Sep
Gallons	145	130	170	180	195	140

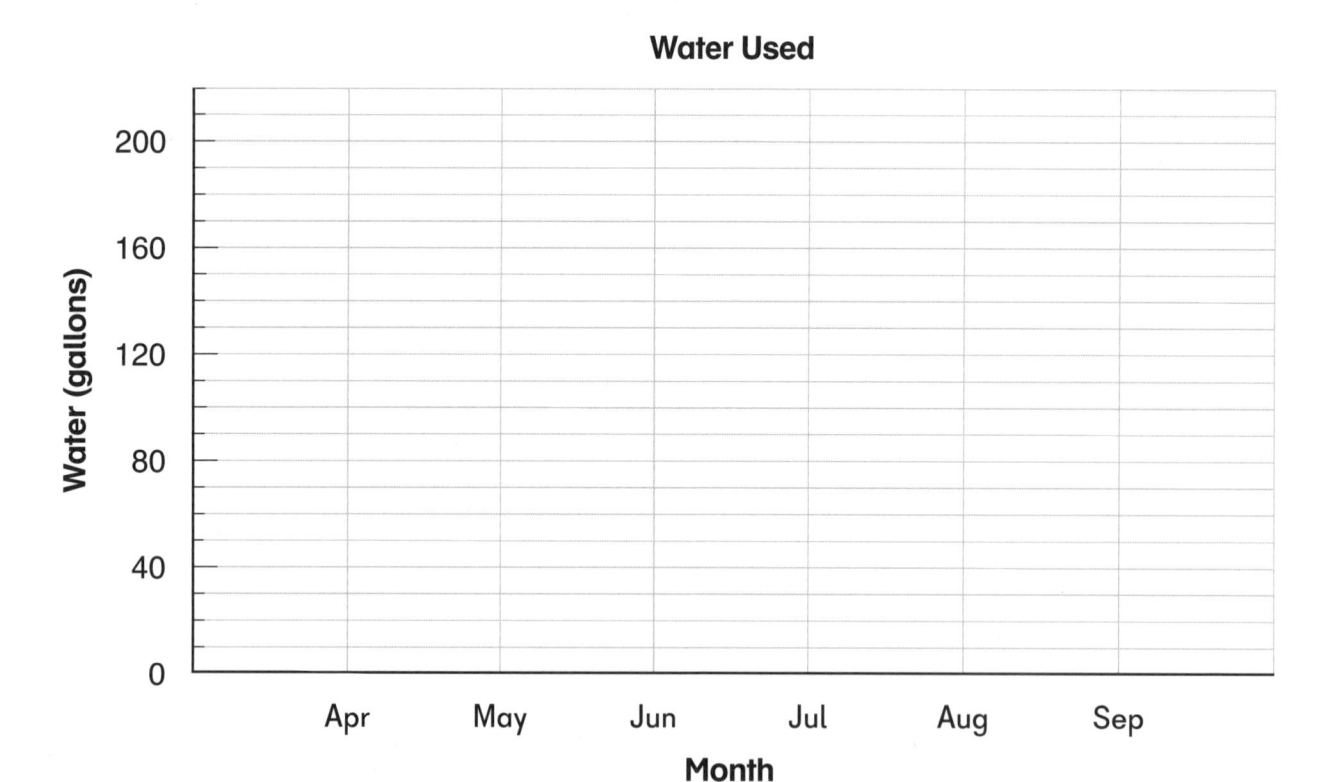

Water Used

(b) What is the total amount of water he used over the 6 months?

(c) What fraction of the months did he use more than 140 gallons? Express the answer in simplest form.

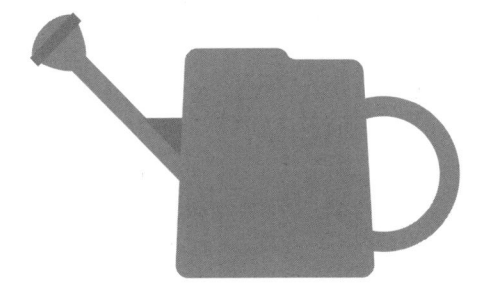

6. Ani baked some cookies. After giving 25 of the cookies to Fuyu and $\frac{2}{5}$ of the remaining cookies to Kawai, she had 48 cookies left. How many cookies did she bake?

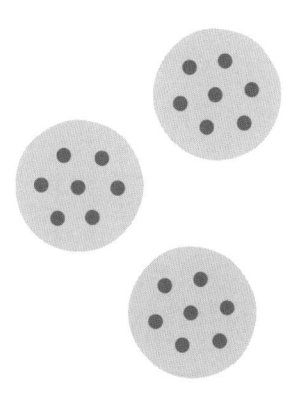

7 1 sheet of paper is 27 cm 8 mm long. How long are 5 sheets of paper laid end to end? Express the answer in meters and centimeters.

8 Nathan had $14.35 less than Parker. Nathan saved another $6.40 and Parker spent $4.65. How much less money does Nathan now have than Parker?

Challenge

9 ABCD is a rectangle. Find the measure of the angles ADE, DEC, and FAB.

| ∠ADE = | ∠DEC = | ∠FAB = |

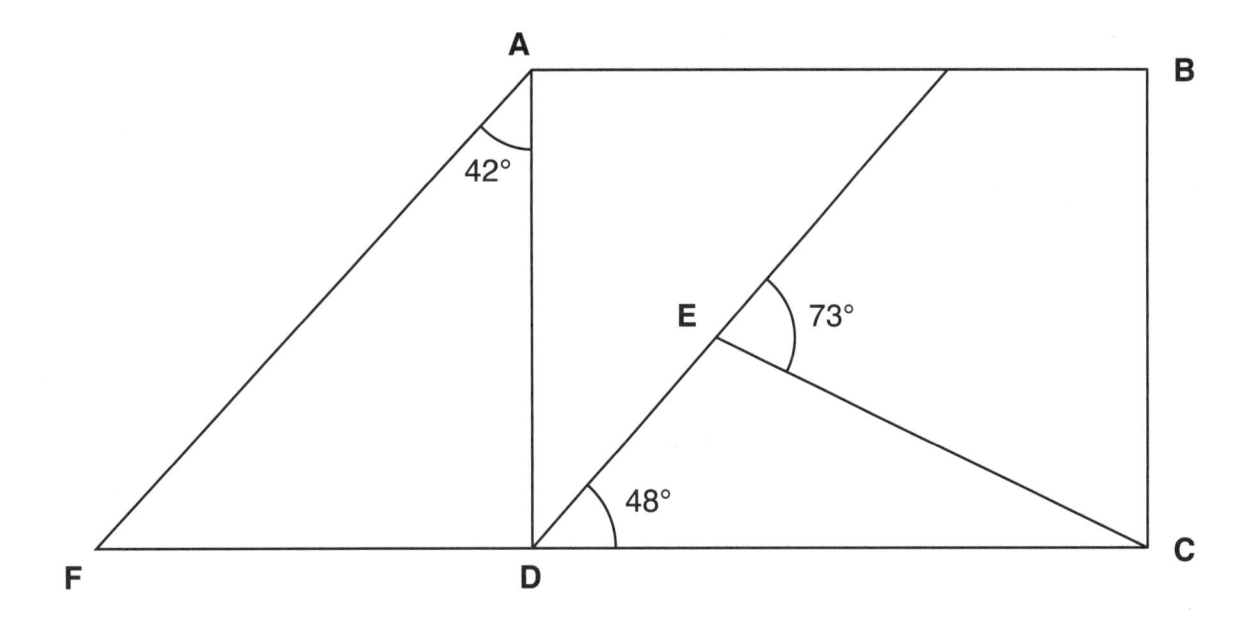

Exercise 6

Check

1 Find the difference between the sum of the first two multiples of 8 and the third multiple of 5.

2 What is the greatest prime number less than 100?

3 30 apples and 45 oranges need to be placed in baskets so that each basket has the same number of fruits. What is the greatest possible number of baskets needed?

4. Mateo was supposed to multiply a number by 7. Instead he divided the number by 7 and got an answer of 8.5. What should the correct answer be?

5. Which of the following have a value greater than 1 but less than 10?

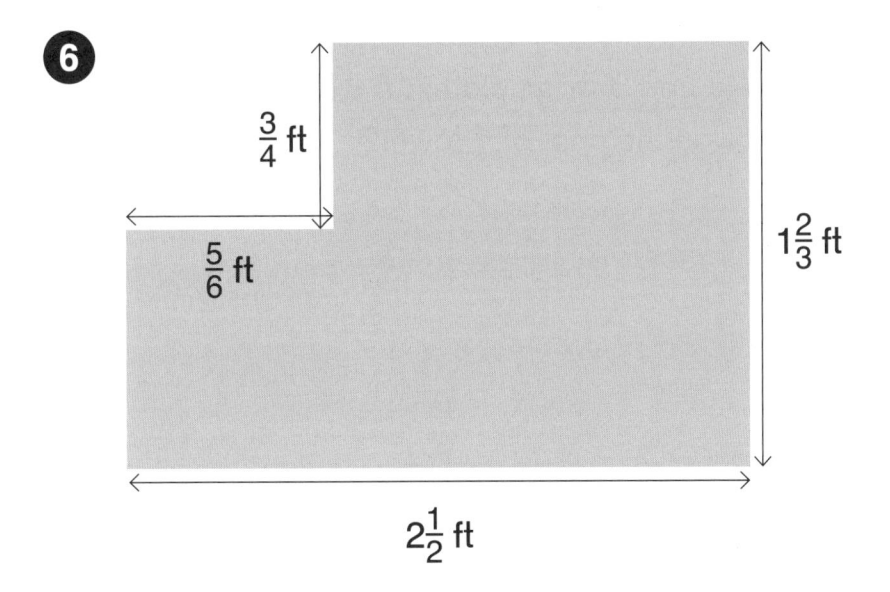

| 0.96 × 9 | 17.8 ÷ 2 | 3.79 × 3 | 12 ÷ 5 | 40.6 ÷ 4 |

6.

$\frac{3}{4}$ ft

$\frac{5}{6}$ ft

$1\frac{2}{3}$ ft

$2\frac{1}{2}$ ft

(a) Express the perimeter of the figure in inches.

(b) Express the area of the figure in square inches.

7. A rope $5\frac{1}{4}$ m long is cut into 5 pieces. 4 of the pieces are $\frac{5}{8}$ m long. How long is the fifth piece? Express the answer in meters as a mixed number in simplest form.

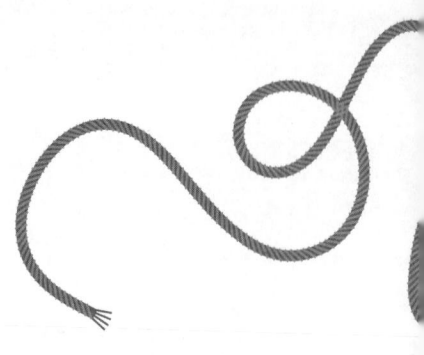

8. Antonio paid $26.85 for 2 towels and 3 wash cloths. A wash cloth cost $2.55 less than a towel. What was the cost of 1 towel?

9 Measure each of the marked angles.

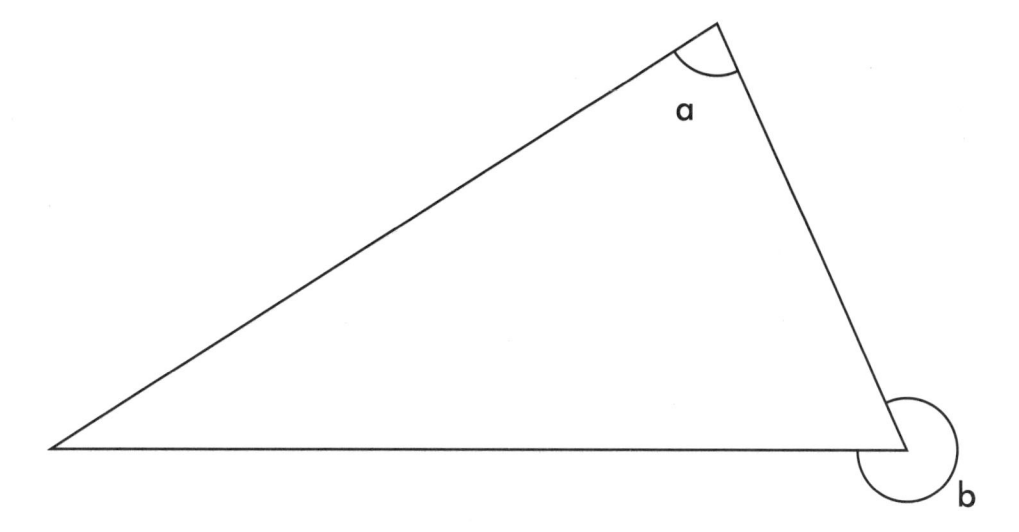

10 (a) Draw an angle that measures 28°.

(b) Draw an angle that measures 280°.

Review 5

11 Complete each figure so that the dotted line is a line of symmetry.

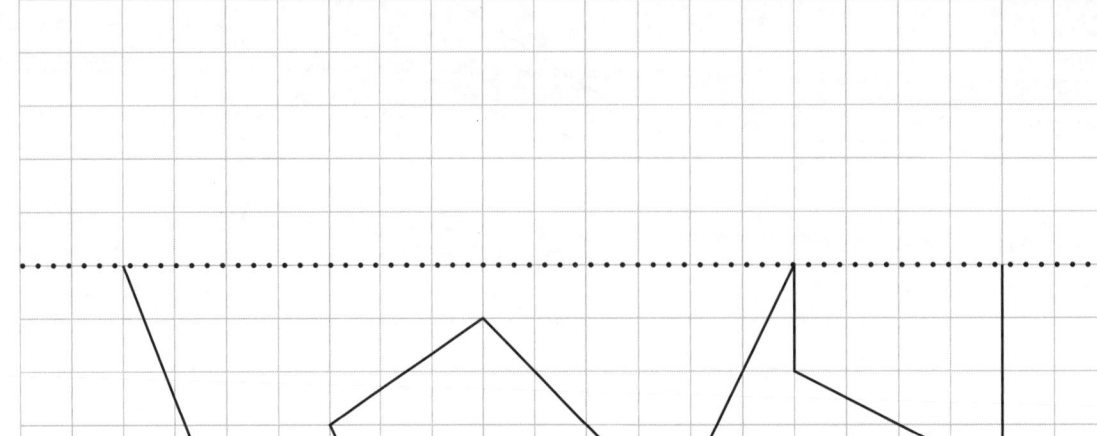

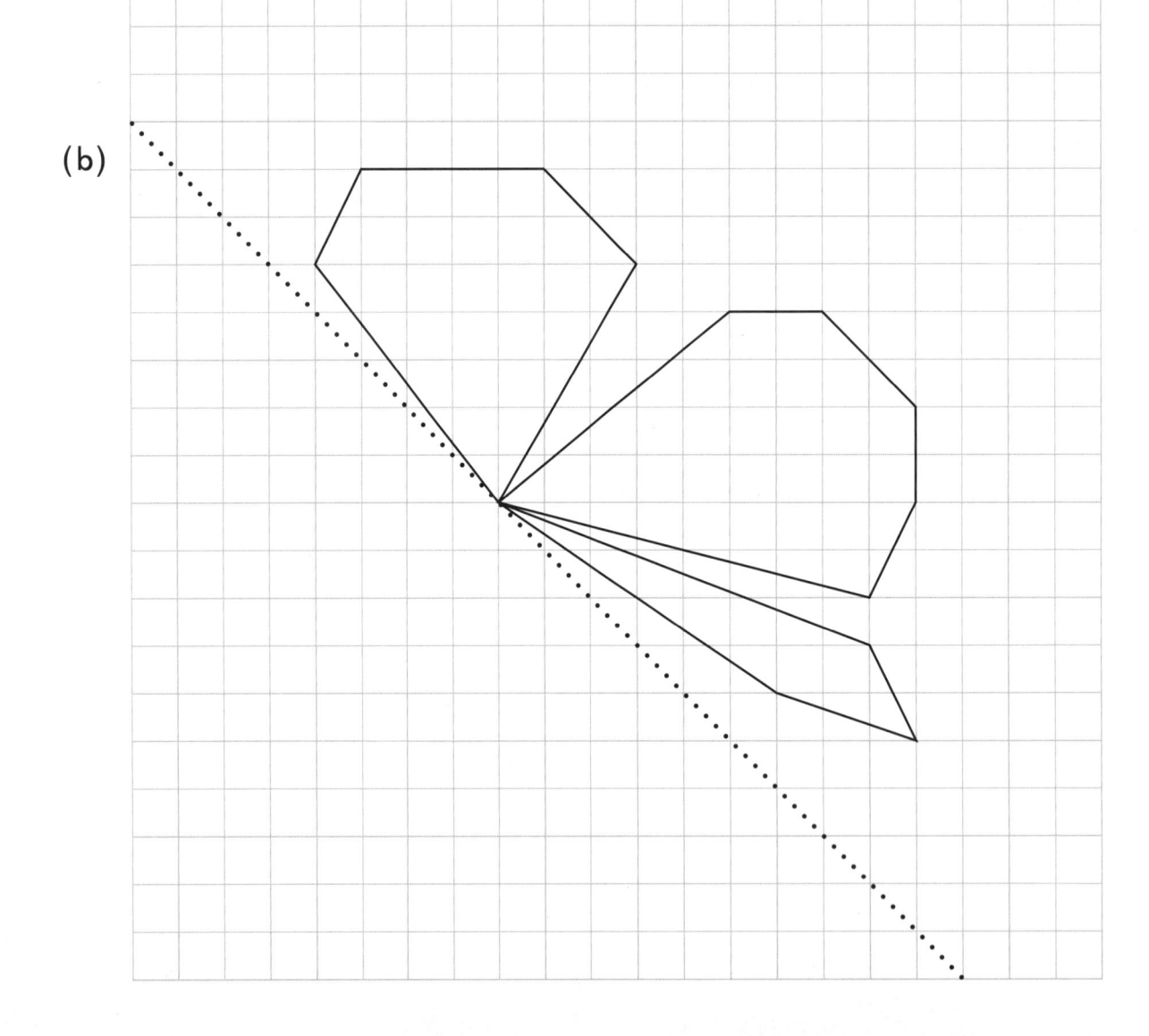

(a)

(b)

Challenge

12 The figure is made up of squares, each with 3-cm sides. What is the perimeter of the entire figure in centimeters?

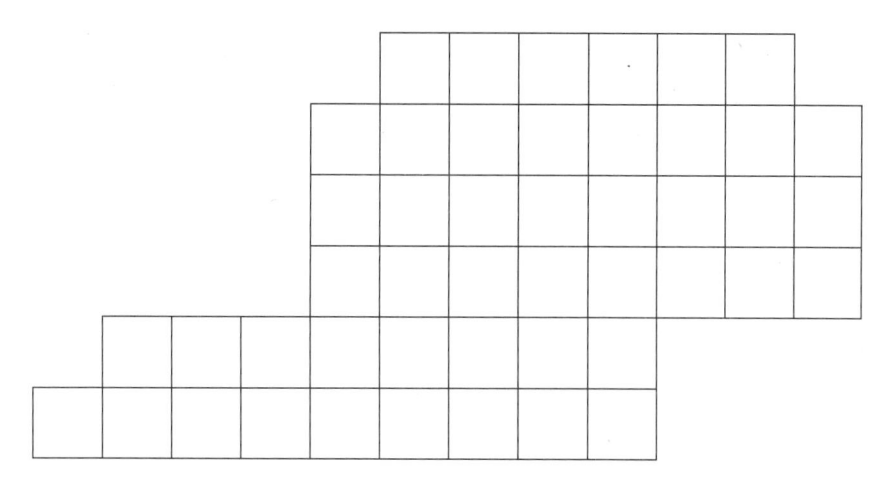

13 The weights of different groups of objects are shown below. How much does the cone weigh?

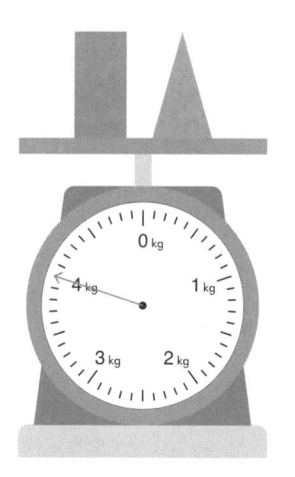

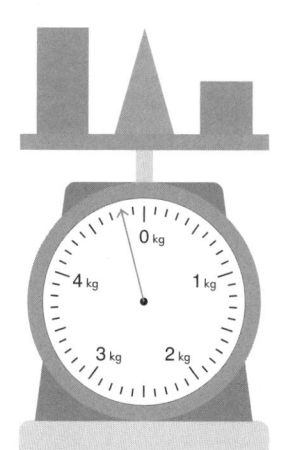

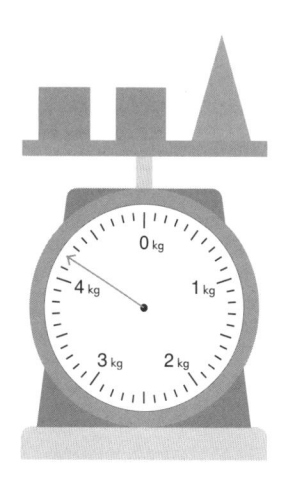

14. The figure shows two different faces of the same cube. In the net of the cube, draw the missing ■ and ○.

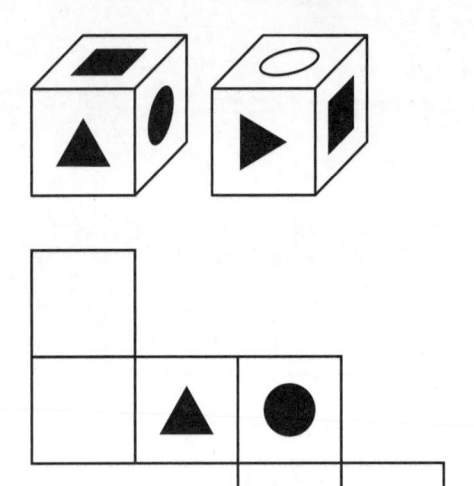

15. This figure is made from three identical overlapping squares. Find the measure of ∠a.

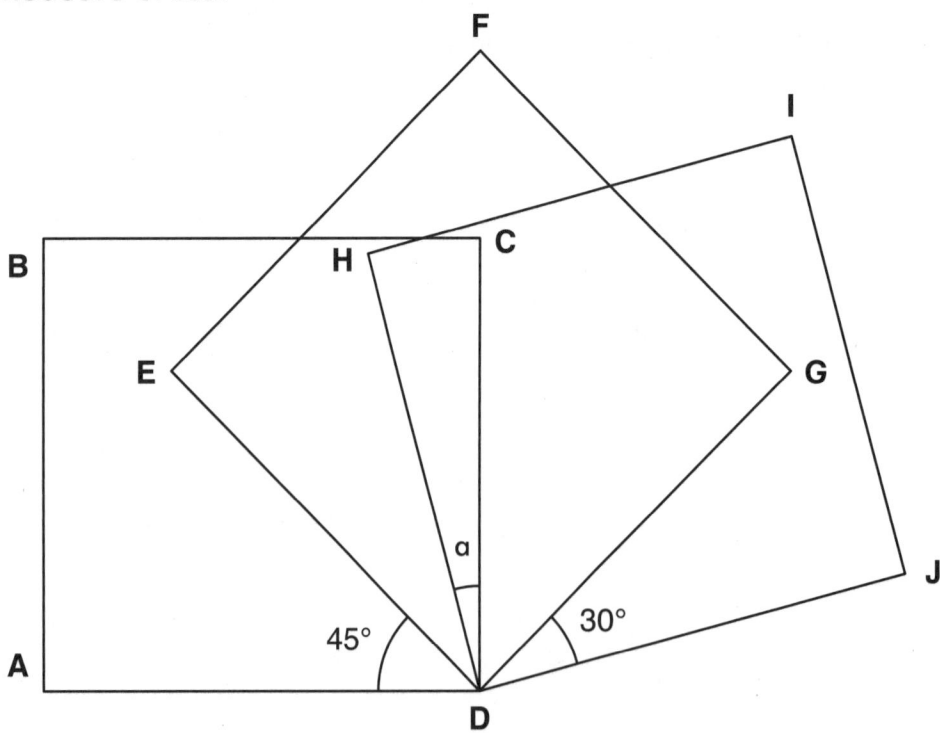

210 Review 5